Photofunctional Nanomaterials and Nanostructures

Photofunctional Nanomaterials and Nanostructures

Guest Editor
Zhixing Gan

 Basel • Beijing • Wuhan • Barcelona • Belgrade • Novi Sad • Cluj • Manchester

Guest Editor
Zhixing Gan
School of Computer and
Electronic Information
Nanjing Normal University
Nanjing
China

Editorial Office
MDPI AG
Grosspeteranlage 5
4052 Basel, Switzerland

This is a reprint of the Special Issue, published open access by the journal *Nanomaterials* (ISSN 2079-4991), freely accessible at: www.mdpi.com/journal/nanomaterials/special_issues/95GAG78V26.

For citation purposes, cite each article independently as indicated on the article page online and using the guide below:

Lastname, A.A.; Lastname, B.B. Article Title. *Journal Name* **Year**, *Volume Number*, Page Range.

ISBN 978-3-7258-3012-1 (Hbk)
ISBN 978-3-7258-3011-4 (PDF)
https://doi.org/10.3390/books978-3-7258-3011-4

© 2025 by the authors. Articles in this book are Open Access and distributed under the Creative Commons Attribution (CC BY) license. The book as a whole is distributed by MDPI under the terms and conditions of the Creative Commons Attribution-NonCommercial-NoDerivs (CC BY-NC-ND) license (https://creativecommons.org/licenses/by-nc-nd/4.0/).

Contents

About the Editor vii

Zhixing Gan
Recent Developments in Photofunctional Nanomaterials and Nanostructures for Emitting, Manipulating, and Harvesting Light
Reprinted from: *Nanomaterials* 2024, 14, 2023, https://doi.org/10.3390/nano14242023 1

Chang-Ruei Li, Yu-Wei Liao, Rashid G. Bikbaev, Jhen-Hong Yang, Lu-Hsing Chen and Dmitrii N. Maksimov et al.
Selective Plasmonic Responses of Chiral Metamirrors
Reprinted from: *Nanomaterials* 2024, 14, 1705, https://doi.org/10.3390/nano14211705 5

Ming Meng, Hucheng Zhou, Jing Yang, Liwei Wang, Honglei Yuan and Yanling Hao et al.
Exploiting the Bragg Mirror Effect of TiO_2 Nanotube Photonic Crystals for Promoting Photoelectrochemical Water Splitting
Reprinted from: *Nanomaterials* 2024, 14, 1695, https://doi.org/10.3390/nano14211695 14

Guangxu Su, Jiangle He, Xiaofei Ye, Hengming Yao, Yaxuan Li and Junzheng Hu et al.
Tailored Triggering of High-Quality Multi-Dimensional Coupled Topological States in Valley Photonic Crystals
Reprinted from: *Nanomaterials* 2024, 14, 885, https://doi.org/10.3390/nano14100885 25

Chang Wang, Xinyu Liu, Yang Zhang, Yan Sun, Zeqing Yu and Zhenrong Zheng
Dual-Channel Switchable Metasurface Filters for Compact Spectral Imaging with Deep Compressive Reconstruction
Reprinted from: *Nanomaterials* 2023, 13, 2854, https://doi.org/10.3390/nano13212854 36

Despoina Tselekidou, Kyparisis Papadopoulos, Konstantinos C. Andrikopoulos, Aikaterini K. Andreopoulou, Joannis K. Kallitsis and Stergios Logothetidis et al.
Optical, Photophysical, and Electroemission Characterization of Blue Emissive Polymers as Active Layer for OLEDs
Reprinted from: *Nanomaterials* 2024, 14, 1623, https://doi.org/10.3390/nano14201623 54

Carlos Tardío, Esther Pinilla-Peñalver, Beatriz Donoso and Iván Torres-Moya
Tunable Unexplored Luminescence in Waveguides Based on D-A-D Benzoselenadiazoles Nanofibers
Reprinted from: *Nanomaterials* 2024, 14, 822, https://doi.org/10.3390/nano14100822 73

Pei Zhang, Yibo Zheng, Linjiao Ren, Shaojun Li, Ming Feng and Qingfang Zhang et al.
The Enhanced Photoluminescence Properties of Carbon Dots Derived from Glucose: The Effect of Natural Oxidation
Reprinted from: *Nanomaterials* 2024, 14, 970, https://doi.org/10.3390/nano14110970 85

Darya Goponenko, Kamila Zhumanova, Sabina Shamarova, Zhuldyz Yelzhanova, Annie Ng and Timur Sh. Atabaev
Hydrophobic and Luminescent Polydimethylsiloxane PDMS-Y_2O_3:Eu^{3+} Coating for Power Enhancement and UV Protection of Si Solar Cells
Reprinted from: *Nanomaterials* 2024, 14, 674, https://doi.org/10.3390/nano14080674 102

Moulay Ahmed Slimani, Jaime A. Benavides-Guerrero, Sylvain G. Cloutier and Ricardo Izquierdo
Enhancing the Performance of Nanocrystalline SnO_2 for Solar Cells through Photonic Curing Using Impedance Spectroscopy Analysis
Reprinted from: *Nanomaterials* **2024**, *14*, 1508, https://doi.org/10.3390/nano14181508 . . . . . . **111**

Larissa Chaperman, Samiha Chaguetmi, Bingbing Deng, Sarra Gam-Derrouich, Sophie Nowak and Fayna Mammeri et al.
Novel Synthesis Route of Plasmonic CuS Quantum Dots as Efficient Co-Catalysts to TiO_2/Ti for Light-Assisted Water Splitting
Reprinted from: *Nanomaterials* **2024**, *14*, 1581, https://doi.org/10.3390/nano14191581 . . . . . . **123**

Junhong Guo, Jinlei Zhang, Yunsong Di and Zhixing Gan
Research Progress on Rashba Effect in Two-Dimensional Organic–Inorganic Hybrid Lead Halide Perovskites
Reprinted from: *Nanomaterials* **2024**, *14*, 683, https://doi.org/10.3390/nano14080683 . . . . **140**

About the Editor

Zhixing Gan

Zhixing Gan obtained his B.S. in Applied Physics in 2010 from Northeastern University, China, and then received his Ph.D. degree at Nanjing University in 2015. After that, he joined Nanjing Normal University as a principal investigator. His primary research interests include photoluminescence, photoresponse, laser manipulation, and photothermal effect of emerging nanomaterials, such as carbon nanomaterials, graphene, 2D materials, and metal halides.

Editorial

Recent Developments in Photofunctional Nanomaterials and Nanostructures for Emitting, Manipulating, and Harvesting Light

Zhixing Gan

Center for Future Optoelectronic Functional Materials, School of Computer and Electronic Information/School of Artificial Intelligence, Nanjing Normal University, Nanjing 210023, China; zxgan@njnu.edu.cn

Citation: Gan, Z. Recent Developments in Photofunctional Nanomaterials and Nanostructures for Emitting, Manipulating, and Harvesting Light. *Nanomaterials* **2024**, *14*, 2023. https://doi.org/10.3390/nano14242023

Received: 2 December 2024
Accepted: 11 December 2024
Published: 16 December 2024

Copyright: © 2024 by the author. Licensee MDPI, Basel, Switzerland. This article is an open access article distributed under the terms and conditions of the Creative Commons Attribution (CC BY) license (https://creativecommons.org/licenses/by/4.0/).

1. Introduction

Photofunctional nanomaterials and nanostructures that can emit, manipulate, convert, and utilize photons in diverse forms have profound meanings, from fundamental understandings to applications. Thus, photofunctional nanomaterials and nanostructures have stimulated trans-disciplinary interests in the fields of physics, chemistry, material science, biology, photons, and engineering while also stimulating scientific breakthroughs in the fields of photovoltaics, photolithography, photoelectronics, photocatalysis, photobiology and phototherapy, photosynthesis, and optical sensing. Recently, photofunctional materials have been developed for storing and processing optical information [1–3]. Neuromorphic optoelectronic devices integrating sensing, storage, and computing abilities have been demonstrated [4–6]. Photofunctional nanomaterials and nanostructures, with their unique appeal, are attracting a growing number of researchers to advance the development of this field.

2. An Overview of Published Articles

This Special Issue brings together eleven articles, including one review article and ten research articles. Four of these articles focus on development of photofunctional metastructures, such as metamirrors, nanotube photonic crystasl, topological photonic devices, and metasurface filters.

Li et al. designed the chiral metamirrors with circular dichroisms of about 0.4 in visible reflection [7]. The chiral metamirrors show high reflectance for right-handed circular polarization with preserved handedness and strongly absorbed left-handed circular polarization on chiroptical resonant wavelengths.

Meng et al. developed a novel bi-layer structure consisting of a top nanotube layer and a bottom nanotube photonic crystal layer in which the photonic bandgap of bottom TiO_2 nanotube photonic crystals can be precisely adjusted by modulating the anodization parameters [8]. The overlapping between the photonic bandgap of photonic crystals with the electronic bandgap of TiO_2 leads to the boosted ultraviolet light absorption of the top TiO_2 layer and enhanced photon-to-current conversion efficiency. This research offers an effective strategy for improving the performance of photoelectrochemical water splitting through intensifying light–matter interactions.

Su et al. computationally proposed a photonic device for the 1550 nm communication band in which topologically protected electromagnetic modes of a high quality can be selectively triggered and modulated on demand [9]. The topological photonic devices can realize Fano lines on the spectrum and show high-quality localized modes by tuning the coupling strength between the zero-dimensional valley corner states and the one-dimensional valley edge states, providing a promising approach for multi-dimensional optical field manipulation in integrated nanophotonic devices.

Wang et al. proposed a compact snapshot compressive spectral-imaging (SCSI) system that leverages the spectral modulations of metasurfaces with dual-channel switchable metasurface filters and employs a deep-learning-based reconstruction algorithm [10]. The proposed SCSI system integrates dual-channel switchable metasurface filters using twisted nematic liquid crystals and anisotropic titanium dioxide nanostructures. The proposed hyperspectral-imaging technology demonstrates superior reconstruction quality and speed compared to those of the traditional compressive hyperspectral image recovery methods. This device is expected to be applied in various areas, such as object detection, face recognition, food safety, biomedical imaging, agriculture surveillance, etc.

The other six research articles focused on the development of photofunctional nanomaterials, with three articles concentrating on light-emitting materials and the final three articles examining light-harvesting materials.

Tselekidou et al. investigated the optical and photophysical characteristics of blue-emitting polymers to promote the understanding of the fundamental mechanisms of color purity and its stability during the operation of Organic Light-Emitting Diode (OLED) devices [11].

Zhang et al. hypothesized that the blue emission of carbon dots (CDs) could be ascribed to the surface states induced by the C–O and C=O groups, while the green luminescence may originate from the deep energy levels associated with the O–C=O groups according to microstructure characterizations, optical measurements, and ultrafiltration experiments [12].

Tardío et al. synthesized and crystallized a set of novel Donor–Acceptor–Donor (D-A-D) benzoselenadiazole derivatives in nanocrystals [13]. The correlation between their chemical structures and the waveguided luminescent properties were explored. The findings revealed that all crystals exhibited luminescence and active optical waveguiding, demonstrating their ability to adjust their luminescence within a broad spectral range of 550–700 nm through suitable chemical functionalization.

Goponenko et al. proposed a novel hydrophobic coating based on a polydimethylsiloxane layer with embedded red-emitting Y_2O_3:Eu^{3+} particles as UV radiation screening and conversion layers for solar cells, resulting in a notable increase in power conversion efficiency by ~9.23% [14]. The developed coating can endure tough environmental conditions, making it potentially useful as a UV-protective, water-repellent, and efficiency-enhancing coating for solar cells.

Slimani et al. studied the intense-pulsed-light-induced crystallization of SnO_2 thin-films using only 500 μs of exposure time [15]. They demonstrated that light-induced crystallization yields improved topography and excellent electrical properties through enhanced charge transfer, improved interfacial morphology, and better ohmic contact compared to thermally annealed SnO_2 films, showing great potential for improved perovskite solar cell manufacturability.

Chaperman synthesized self-doped CuS nanoparticles via a microwave-assisted polyol process to act as co-catalysts to TiO_2 nanofiber-based photoanodes for visible light-assisted water electrolysis [16]. These low-cost and easy-to-achieve composite materials allow for an improved overall efficiency of water oxidation (and consequently hydrogen generation at the Pt counter electrode) in passive electrolytes, even with a 0 V bias.

Furthermore, in the mini-review, Guo et al. summarized recent research progress on the Rashba effect of two-dimensional (2D) organic–inorganic hybrid perovskites [17]. The origin and magnitude of Rashba spin splitting, the layer-dependent Rashba band splitting of 2D perovskites, and the Rashba effect of different 2D perovskites are discussed. Moreover, related applications in regard to photodetectors and photovoltaics are reviewed. Future research to modulate the Rashba strength is expected to promote the optoelectronic and spintronic applications of 2D perovskites.

3. Conclusions

In summary, this Special Issue mainly reports recent research progress regarding photofunctional nanomaterials and nanostructures that emit, manipulate, and harvest light. These contributions are expected to promote the development of integrated nanophotonic chips, hyperspectral imaging, photoelectrochemical water splitting, solar cells, and OLEDs. We hope that this Special Issue will help readers to gain more insight into this area and also provide helpful guidance for the future development of photofunctional nanomaterials and nanostructures.

Funding: This work was supported by the Natural Science Foundation of Shandong Province (ZR2021YQ32), the Taishan Scholars Program of Shandong Province (tsqn201909117), Opening Foundation of Hubei Key Laboratory of Photoelectric Materials and Devices (PMD202401), and the special fund for Science and Technology Innovation Teams of Shanxi Province.

Conflicts of Interest: The authors declare no conflict of interest.

References

1. Sheng, Y.; Gui, Q.; Zhang, Y.; Yang, X.; Xing, F.; Liu, C.; Di, Y.; Wei, S.; Cao, G.; Yang, X.; et al. Lead-Free Perovskites with Photochromism and Reversed Thermochromism for Repeatable Information Writing and Erasing. *Adv. Funct. Mater.* **2024**, *34*, 2406995. [CrossRef]
2. Zhang, M.; Yang, L.; Wu, X.; Wang, J. Black Phosphorus for Photonic Integrated Circuits. *Research* **2023**, *6*, 0206. [CrossRef] [PubMed]
3. Tan, D.; Wang, Z.; Xu, B.; Qiu, J. Photonic circuits written by femtosecond laser in glass: Improved fabrication and recent progress in photonic devices. *Adv. Photonics* **2021**, *3*, 024002. [CrossRef]
4. Li, M.; Li, C.; Ye, K.; Xu, Y.; Song, W.; Liu, C.; Xing, F.; Cao, G.; Wei, S.; Chen, Z.; et al. Self-Powered Photonic Synapses with Rapid Optical Erasing Ability for Neuromorphic Visual Perception. *Research* **2024**, *7*, 0526. [CrossRef]
5. Wang, Y.; Zha, Y.; Bao, C.; Hu, F.; Di, Y.; Liu, C.; Xing, F.; Xu, X.; Wen, X.; Gan, Z.; et al. Monolithic 2D Perovskites Enabled Artificial Photonic Synapses for Neuromorphic Vision Sensors. *Adv. Mater.* **2024**, *36*, 2311524. [CrossRef]
6. Li, T.; Miao, J.; Fu, X.; Song, B.; Cai, B.; Ge, X.; Zhou, X.; Zhou, P.; Wang, X.; Jariwala, D.; et al. Reconfigurable, non-volatile neuromorphic photovoltaics. *Nat. Nanotechnol.* **2023**, *18*, 1303–1310. [CrossRef] [PubMed]
7. Li, C.-R.; Liao, Y.-W.; Bikbaev, R.G.; Yang, J.-H.; Chen, L.-H.; Maksimov, D.N.; Pankin, P.S.; Timofeev, I.V.; Chen, K.-P. Selective Plasmonic Responses of Chiral Metamirrors. *Nanomaterials* **2024**, *14*, 1705. [CrossRef] [PubMed]
8. Meng, M.; Zhou, H.; Yang, J.; Wang, L.; Yuan, H.; Hao, Y.; Gan, Z. Exploiting the Bragg Mirror Effect of TiO_2 Nanotube Photonic Crystals for Promoting Photoelectrochemical Water Splitting. *Nanomaterials* **2024**, *14*, 1695. [CrossRef] [PubMed]
9. Su, G.; He, J.; Ye, X.; Yao, H.; Li, Y.; Hu, J.; Lu, M.; Zhan, P.; Liu, F. Tailored Triggering of High-Quality Multi-Dimensional Coupled Topological States in Valley Photonic Crystals. *Nanomaterials* **2024**, *14*, 885. [CrossRef] [PubMed]
10. Wang, C.; Liu, X.; Zhang, Y.; Sun, Y.; Yu, Z.; Zheng, Z. Dual-Channel Switchable Metasurface Filters for Compact Spectral Imaging with Deep Compressive Reconstruction. *Nanomaterials* **2023**, *13*, 2854. [CrossRef]
11. Tselekidou, D.; Papadopoulos, K.; Andrikopoulos, K.C.; Andreopoulou, A.K.; Kallitsis, J.K.; Logothetidis, S.; Laskarakis, A.; Gioti, M. Optical, Photophysical, and Electroemission Characterization of Blue Emissive Polymers as Active Layer for OLEDs. *Nanomaterials* **2024**, *14*, 1623. [CrossRef] [PubMed]
12. Zhang, P.; Zheng, Y.; Ren, L.; Li, S.; Feng, M.; Zhang, Q.; Qi, R.; Qin, Z.; Zhang, J.; Jiang, L. The Enhanced Photoluminescence Properties of Carbon Dots Derived from Glucose: The Effect of Natural Oxidation. *Nanomaterials* **2024**, *14*, 970. [CrossRef] [PubMed]
13. Tardío, C.; Pinilla-Peñalver, E.; Donoso, B.; Torres-Moya, I. Tunable Unexplored Luminescence in Waveguides Based on D-A-D Benzoselenadiazoles Nanofibers. *Nanomaterials* **2024**, *14*, 822. [CrossRef]
14. Goponenko, D.; Zhumanova, K.; Shamarova, S.; Yelzhanova, Z.; Ng, A.; Atabaev, T.S. Hydrophobic and Luminescent Polydimethylsiloxane PDMS-Y_2O_3:Eu^{3+} Coating for Power Enhancement and UV Protection of Si Solar Cells. *Nanomaterials* **2024**, *14*, 674. [CrossRef] [PubMed]
15. Slimani, M.A.; Benavides-Guerrero, J.A.; Izquierdo, S.G.C.R. Enhancing the Performance of Nanocrystalline SnO_2 for Solar Cells through Photonic Curing Using Impedance Spectroscopy Analysis. *Nanomaterials* **2024**, *14*, 1508. [CrossRef] [PubMed]

16. Chaperman, L.; Chaguetmi, S.; Deng, B.; Gam-Derrouich, S.; Nowak, S.; Mammeri, F.; Ammar, S. Novel Synthesis Route of Plasmonic CuS Quantum Dots as Efficient Co-Catalysts to TiO_2/Ti for Light-Assisted Water Splitting. *Nanomaterials* **2024**, *14*, 1581. [CrossRef] [PubMed]
17. Guo, J.; Zhang, J.; Di, Y.; Gan, Z. Research Progress on Rashba Effect in Two-Dimensional Organic–Inorganic Hybrid Lead Halide Perovskites. *Nanomaterials* **2024**, *14*, 683. [CrossRef] [PubMed]

Disclaimer/Publisher's Note: The statements, opinions and data contained in all publications are solely those of the individual author(s) and contributor(s) and not of MDPI and/or the editor(s). MDPI and/or the editor(s) disclaim responsibility for any injury to people or property resulting from any ideas, methods, instructions or products referred to in the content.

Article

Selective Plasmonic Responses of Chiral Metamirrors

Chang-Ruei Li [1], Yu-Wei Liao [2], Rashid G. Bikbaev [3,4], Jhen-Hong Yang [2], Lu-Hsing Chen [5], Dmitrii N. Maksimov [3,4], Pavel S. Pankin [3,4], Ivan V. Timofeev [3,4] and Kuo-Ping Chen [5,6,*]

[1] Institute of Lighting and Energy Photonics, College of Photonics, National Yang Ming Chiao Tung University, 301 Gaofa 3rd Road, Tainan 71150, Taiwan; e1253400@gmail.com
[2] Institute of Photonic System, College of Photonics, National Yang Ming Chiao Tung University, 301 Gaofa 3rd Road, Tainan 71150, Taiwan; yuwei880404@gmail.com (Y.-W.L.); s87069@hotmail.com (J.-H.Y.)
[3] Kirensky Institute of Physics, Federal Research Center KSC SB RAS, Krasnoyarsk 660036, Russia; batblr_90@mail.ru (R.G.B.); mdn@tnp.krasn.ru (D.N.M.); p.s.pankin@mail.ru (P.S.P.); ivan-v-timofeev@ya.ru (I.V.T.)
[4] Institute of Engineering Physics and Radioelectronics, Siberian Federal University, Krasnoyarsk 660041, Russia
[5] Institute of Photonics Technologies, National Tsing Hua University, Hsinchu 30013, Taiwan; wilson20000426@yahoo.com.tw
[6] Institute of Imaging and Biomedical Photonics, College of Photonics, National Yang Ming Chiao Tung University, 301 Gaofa 3rd Road, Tainan 71150, Taiwan
* Correspondence: kpchen@ee.nthu.edu.tw

Abstract: The properties of circularly polarized light has recently been used to selectively reflect chiral metasurfaces. Here we report the more complete basic functionalities of reflectors and absorbers that display various optical phenomena under circularly polarized light at normal incidence as before. For the chiral metamirrors we designed, the circular dichroism in about 0.4 reflection is experimentally observed in visible wavelengths. The experimental results also show high reflectance for right-handed circular polarization with preserved handedness and strongly absorbed left-handed circular polarization at chiroptical resonant wavelengths. By combining a nanobrick and wire grating for our design, we find and offer a new structure to demonstrate the superposition concept of the phase in the same plane that is helpful in effectively designing chiral metamirrors, and could advance development of their ultracompact optical components.

Keywords: metamirrors; circular dichroism; chirality; visible wavelengths

1. Introduction

The term chirality is identified as the property of an object that lacks any mirror symmetry plane [1–7] and is a fundamental characteristic of natural molecules and artificial metamaterials which have different optical properties for right-handed circular polarization (RCP) or left-handed circular polarization (LCP). A famous example in nature, the Chrysina gloriosa, a jeweled scarab beetle, can selectively reflect left-handed circularly polarized light in reflection, because of the particular textures of its exoskeleton [8]. Moreover, this chiroptical response is utilized by Chrysina gloriosa to perceive and communicate with its companions [9]. However, the signal of chiroptical response is universally weaker in natural as compared to artificial metamaterial. Therefore, in recent years the chiral metasurface [10–13] has been widely studied to achieve a strong chiroptical response in applications such as circular polarizers [14–18], hot-electron collection devices [19–23], optical encryption [24,25], and biosensors for analyzing circular dichroism spectroscopy [26–30].

In this paper, we demonstrate all basic functionalities for a series of reflectors and absorbers consisting of a metasurface on the top of a conventional mirror under circularly polarized light at normal incidence (as shown in Figure 1) that is more complete than before [31,32]. A conventional isotropic mirror can reverse the handedness of circularly polarized when light is reflected off its surface, because it can reverse the propagation direction of the reflected electromagnetic wave of the electric field (seen in Figure 1a), as has previously been discussed in detail [33,34]. The necessary condition for selective reflection without preserving handedness for both LCP absorbers and RCP absorbers under circular polarized light is only to break mirror symmetry to the perpendicular light propagation direction of the pattern plane, illustrated in Figure 1b,c. In addition, Figure 1b,c are opposite phenomena and mirror images of each other. Figure 1d shows an anisotropy mirror designed as a half-wave plate with a phase difference of π [35–38] and having the property of preserving handedness without handedness variation. The LCP and RCP mirrors are designed to not only selectively reflect one circular polarized light while preserving handedness, but also totally absorb the other, as shown in Figure 1e,f [30]. In addition, a linear polarization perfect absorber composed of a metasurface and a thick backplane of a general mirror separated by a thin lossy dielectric spacer, has been reported [39,40]. It is important to note that an isotropic linear polarization absorber can absorb both linear polarizations [41–43], and is the circular polarization perfect absorber to absorb both RCP and LCP light, as shown Figure 1g. The mirrors shown in Figure 1 allow for control of the polarization and intensity of the reflected light, which is essential when designing various devices. This study aims to develop a metamirror with the maximum value of circular dichroism. The metamirror analysis revealed that circular dichroism can only be obtained by using LCP or RCP mirrors (see Figure 1e,f). For this reason, in this work we present a simple approach that utilizes the superposition concept of the phase to design chiral metamirrors from a single structure layer, by combining the nanobrick and the wire grating in the same plane. Furthermore, this single patterned layer can make a good contribution to lowering the complexity of the fabricated device and its cost. The analysis of both simulation and experiments with reflectance coefficients spectra demonstrates the chiral metamirrors we have designed can selectively reflect the RCP light while preserving the handedness and absorbing the LCP light. In addition, the simulated results display good agreement with the experimental results showing reflectance spectra and spectral shape.

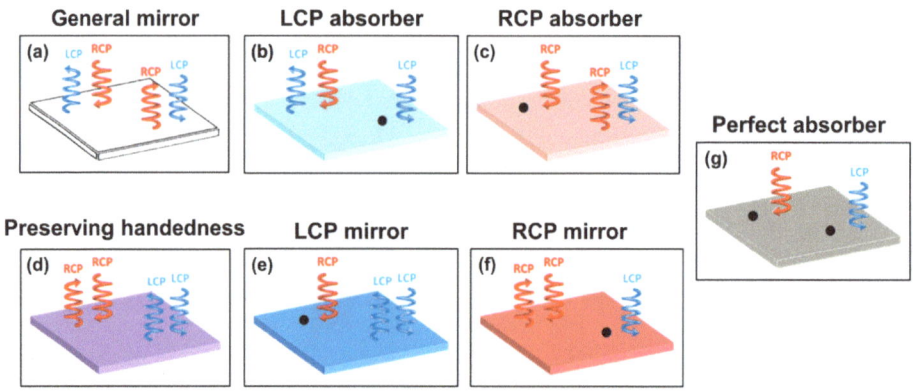

Figure 1. Schematic of all reflectors under circularly polarized light at normal incidence. (**a**) A general mirror reverses the handedness of circular polarized light in reflection. (**b**) An LCP absorber and (**c**) an RCP absorber can reflect circularly polarized light of one handedness with handedness variation, while absorbing the other handedness, as shown as the black circle. (**d**) An anisotropy mirror is designed as a half-wave plate with a phase difference of π, to preserve handedness without handedness variation. (**e**) An LCP mirror and (**f**) an RCP mirror can reflect circular polarized light of one handedness and preserve handedness without handedness variation, while absorbing the other handedness. (**g**) An isotropic mirror absorbs both LCP and RCP light and does not reflect light as a perfect absorber.

2. Materials and Methods

In order to design the circular dichroism metamirrors, conditions including breaking mirror symmetry and the *n*-fold (*n* > 2) rotational symmetry [8] must be satisfied. Therefore, we combined the nanobrick and the wire grating in the same plane. Figure 2 shows a three-dimensional illustration of the gold nanobrick–wire grating complex structure. The schematic configuration is designed as a 60 nm gold structure on top of a 130 nm silicon dioxide spacer, underneath which is a layer of thick gold reflector and a glass substrate. Due to the optically thick gold reflector in this design, transmission can be neglected; the chiroptical response can be tremendously enhanced by providing a resonant cavity. The width of both the nanobrick and the wire grating was $g = l = 70$ nm, and the length of the nanobrick was $w = 190$ nm. The unit cell was replicated in a two-dimensional square lattice along the x and y axes with period $P_x = P_y = P = 250$ nm. The nanobrick was rotated by 45 degrees with respect to the z-axis to break the mirror symmetry and the *n*-fold (*n* > 2) rotational symmetry. It is worth noting that we demonstrate that the gold nanobrick-wire grating complex structure of the chiral metamirrors planar photonic structures with a single patterned layer not only produces a chiral optical response, but also lowers the complexity of a fabricated device and the cost.

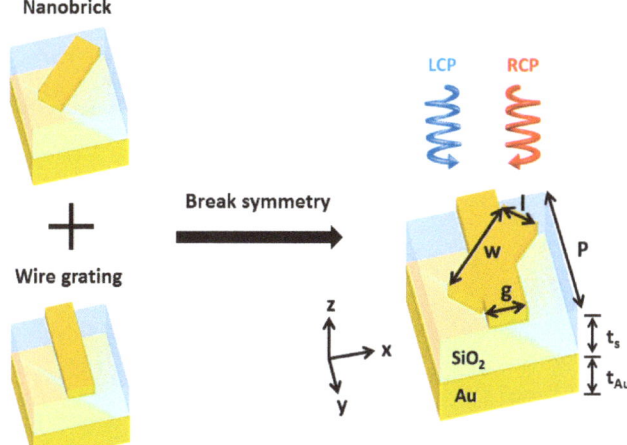

Figure 2. Schematic configuration of the chiral metamirrors consisting of the nanobrick and the wire grating in the same plane for breaking mirror symmetry and the *n*-fold (*n* > 2) rotational symmetry. A unit cell of the metasurface consists of a gold structure which is separated from a thick gold reflector by a thin SiO$_2$ spacer. Geometrical parameters: $P = 250$ nm, $g = l = 70$ nm, $w = 190$ nm, $t_s = 130$ nm, $t_{Au} = 200$ nm.

3. Results

3.1. Simulated Results

Simulated results of the reflectance spectra under RCP and LCP incident light based on the three-dimensional Finite-Difference Time-Domain method are illustrated in Figure 3a,b. To better describe the observed chiroptical response, the corresponding circular dichroism is illustrated in Figure 3a. The circular dichroism spectrum was characterized by different absorption between RCP and LCP light (CD = $A_{LCP} - A_{RCP}$). The total reflectance spectra indicated clearly different reflection under RCP and LCP incident light, and the circular dichroism was ~0.48 at a wavelength of 640 nm. In contrast to the phenomenon of the RCP light, which was reflected, the LCP was absorbed by our chiral metamirrors. Moreover, the corresponding reflecting behavior of the reflectance coefficients of the co-polarization and cross-polarization components of the chiral metamirrors under RCP and LCP incident light is depicted in Figure 3b. The reflectance coefficient $|r_{RR}|^2$ ($|r_{LR}|^2$) is defined as the RCP (LCP) light reflected off the surface of the chiral metamirrors under RCP incident light, while $|r_{RL}|^2$ ($|r_{LR}|^2$) indicates the cross polarization of circularly polarized light. As expected, $|r_{RL}|^2$ and $|r_{LR}|^2$ were exactly the same because of the symmetry of the unit cell [44]. In addition, Figure 3b demonstrates our chiral metamirrors not only selectively reflected the RCP light, but also preserved the handedness at the chiroptical resonance wavelength of 640 nm, while all LCP components were completely absorbed. In order to preserve handedness, there is a need to design the anisotropy structure with a phase difference of π as a half-wave plate. For our designed dichroism metamirrors, the superposition concept of the phase is illustrated in Figure 3c. First, we modified two different lengths of brick to create two kinds of phase difference. Then, we combined these two structures to superpose the phase to create the phase difference of π, to preserve the handedness that is one of the necessary conditions for designing chiral metamirrors. Consequently, we have successfully demonstrated this simple way to design chiral metamirrors which produce the chiroptical response while preserving handedness by combining the nanobrick and the wire grating in the same plane. The chiral selective reflectance spectra demonstrated the distinct resonance modes under the RCP and LCP incident light. To better describe the fundamental physical mechanism of the chiral resonance at a wavelength of 640 nm while the value of circular dichroism is maximal, the cuts in the electric field distribution through the middle of the gold metasurface were as shown in Figure 3d,e. In the near field distribution, the electric field localizes to a different position, which means the resonance mode is different, resulting in different reflectance spectra. Furthermore, it is evident that the stronger field restriction on both sides of long axis of the nanobrick corresponded to the lower reflectance under LCP light.

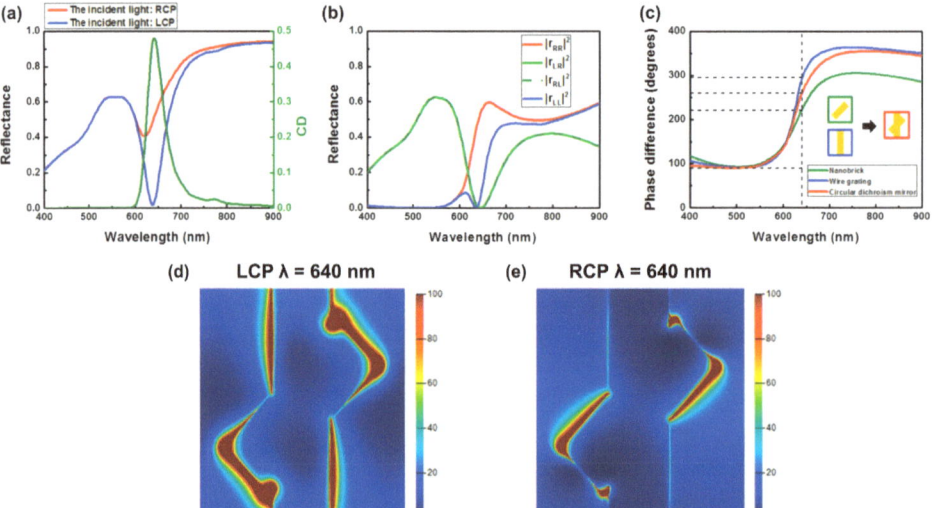

Figure 3. Simulated results of the chiral metamirrors. (**a**) Total reflectance spectra and (**b**) reflectance coefficients of the co-polarization and cross-polarization components of the chiral metamirrors under RCP and LCP incident light. (**c**) Phase difference of reflected light for the three types of structure under circularly polarized light. Electric field distributions for the chiral metamirrors at the maximum chiroptical response wavelength of 640 nm under (**d**) LCP and (**e**) RCP incident light.

In principle, the circular polarizations can be decomposed into two linear polarizations with orthogonal direction in a 90° phase shift. The anisotropy of our designed structure can produce the phase difference and the transformation of the linear polarization state into its orthogonal one. This phase difference will lead to destructive interference and constructive interference for linear polarizations under LCP and RCP light, respectively. The ideal conditions of the phase and the amplitude for maximizing circular dichroism (maximizing LCP absorption, minimizing RCP absorption) in the planar chiral metamaterial are as follows [19]:

$$\phi_{xx} + 90° = \phi_{xy} = \phi_{yx} = \phi_{yy} + 270° \tag{1}$$

$$|r_{xx}| = |r_{yy}| = |r_{xy}| = |r_{yx}| \tag{2}$$

As mentioned above, our chiral metamirrors have also satisfied these two conditions in achieving the maximizing circular dichroism at the resonant wavelength of 640 nm as shown in Figure 4. As shown in Equation (2) and Figure 4b, the same amplitude will not only cancel out the same linear polarizations of the reflected light for destructive interference, but also enhance the intensity of the other polarization for constructive interference. A more detailed description of the process for achieving the required phase difference and producing a half-wave plate metasurface has been presented in our previous work [38].

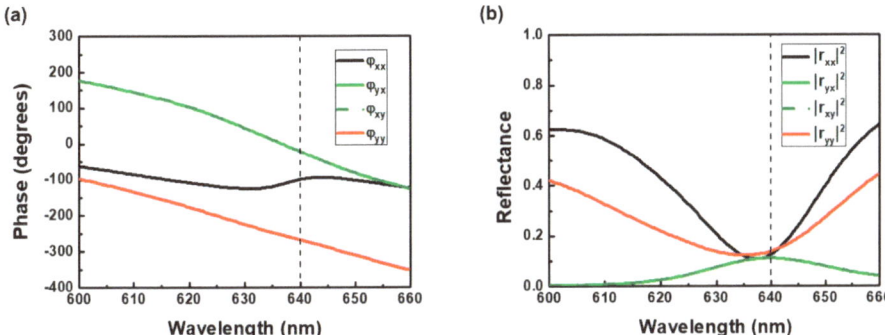

Figure 4. Simulated results of the chiral metamirrors. (**a**) Phase and (**b**) amplitude of reflectance spectra under linearly polarized light.

3.2. Experimental Results

To experimentally realize the chiral metamirrors, we demonstrate a series of experiments in the visible wavelengths. E-beam lithography was used to fabricate the metasurface. First, the 200 nm gold film and 130 nm amorphous SiO$_2$ were deposited via sputter on the glass substrate. The PMMA A4 photoresist with a thickness equal to 200 nm was spin coated on the substrate at 1500 rpm for 15 s and 8000 rpm for 30 s, respectively. Then, it was prebaked at 180 °C for 90 s. The lower spin speed was for flatness, and higher spin speed for controlling thickness. The E-beam lithography system (ELS−7500EX, ELIONIX, Wellesley, MA, USA) was used to define the pattern, and the photoresist was developed using developers (MIBK:IPA = 1: 3) for 30 s. Then, 60 nm gold film was deposited using the electron gun evaporation system (VT1-10CE, ULVAC, Munich, Germany) with a 0.5 Å/s deposition rate in 5 × 10^{-6} Torr. The sample was immersed in acetone liquid to lift off residual photoresist and unnecessary gold film. During the gold deposition process, we encountered an issue with its adhesion to the SiO$_2$ surface. To address this, we deposited a thin 5 nm titanium film on the SiO$_2$ prior to applying the gold. The scanning electron microscope image of the fabricated structure is shown in Figure 5a. Figure 5c shows the measured total reflectance spectra at wavelengths between 400 nm and 900 nm under RCP and LCP incident light. A comparison of the experimental spectra with the theoretical ones (see Figure 3a) reveals that the Q-factor of the resonance in the actual experiment is significantly lower than in the simulation. This significant decrease in Q-factor is due to the inclusion of a thin titanium layer, which was necessary to enhance the adhesion between gold and SiO$_2$. The simulated reflection spectra for the structure with a thin titanium film under RCP and LCP incident light are shown in Figure 5b. As expected, a clearly chiral selective reflection appears, as shown in Figure 5b,c. The minimum reflectance for LCP waves reached ~5% at the chiroptical resonant wavelength in the measurement. However, by comparing the simulated and experimental results we can observe the reflectance spectra under circular polarized light underwent a blue shift because of the wider width of the wire grating of ~15 nm in the experiment. Nevertheless, the experimental results match well with the simulated results. Figure 5e,f plot the reflectance coefficients of the co-polarization and cross-polarization components of the chiral metamirrors under RCP and LCP incident light, fully providing information on the chiroptical response. We observed the RCP light can be reflected with preserved handedness on our chiral metamirrors at the resonant wavelength of 670 nm, and the LCP components were absorbed almost simultaneously, as shown in Figure 5f. Nonetheless, the reflectance coefficients spectra also underwent a slight blue shift because of the inaccuracy from the sample fabrication, as mentioned above, and show a slight imprecision due to the optical components at longer than 800 nm.

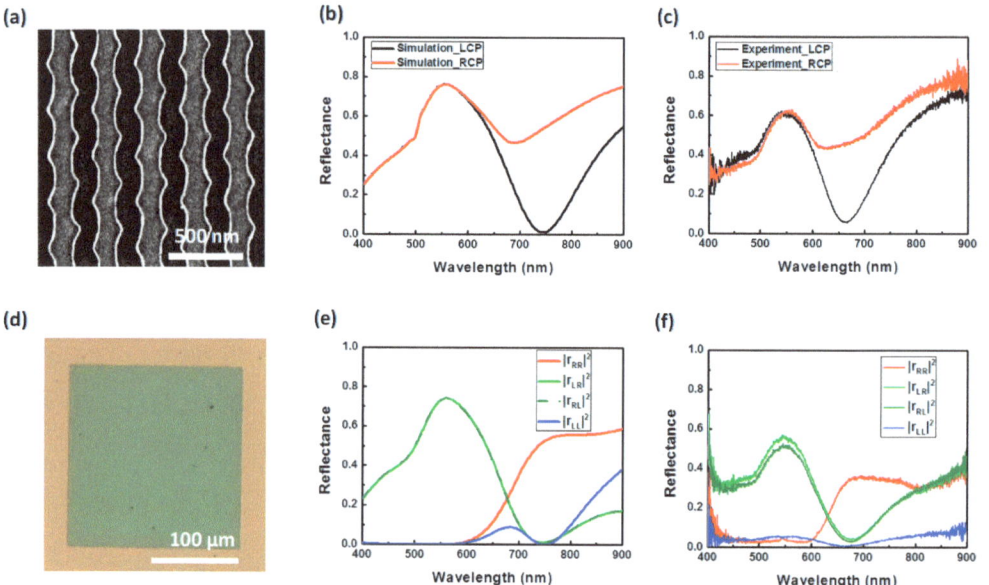

Figure 5. Experimentally measured results of optical reflectance spectra of the chiral metamirrors with a thin adhesion layer of titanium. (**a**) Scanning electron microscopy image of the fabricated sample of the chiral metamirrors. The scale bar is 500 nm. (**b**) Simulated and (**c**) experimental results of the total reflectance spectra under LCP (black) and RCP (red) incident light. (**d**) Optical microscopy images of the metamirrors show good uniformity in e-beam lithography. The scale bar is 100 μm. (**e**) Simulated and (**f**) experimental results of reflectance coefficients of the co-polarization and cross-polarization components of the chiral metamirrors under RCP and LCP incident light.

4. Discussion

In summary, we have presented more complete basic functionalities of reflectors and absorbers than previously demonstrated, and an approach to the superposition concept of the phase in the same plane that is helpful in designing chiral metamirrors. We have also demonstrated the design of our metamirrors offers significant magnitudes of chiroptical response in both simulated and experimental results in the visible wavelengths. The theoretical reflectance is nearly absorbed at the chiroptical resonant wavelength under LCP incident light, while a reflectance of ~5% is experimentally proven. We believe that the high resolution for chiral selection in reflection is an attractive feature for many applications in optical components such as circular polarizers and absorbed filters. In addition, applying the characteristic of preserving handedness to circular polarizers could avoid transforming one polarization state into the other. This transformation can complicate an optical system and influence the measurement of samples in ways we do not expect. It is important to note that our designed single-patterned layer is a useful contribution to decreasing the complexity of fabrication requirements such as the high cost of a multilayered patterned layer [8] and the accurate alignment, and can also achieve a highly efficient chiroptical response. Finally, the proposed structure, consisting of nanowires coupled with nanobricks, can be used in the design of beam steering devices [45,46]. As in previous studies, the nanowires do not connect, allowing them to be used as electrical contacts and, consequently, to change the diffraction grating period. The advantage of this proposed chiral metamirror in contrast to conventional nanostrips is that it allows for the creation of a diffraction grating. The proposed metamirror allows control of the diffraction angles, as well as the intensity of diffraction orders and the polarization.

Author Contributions: Conceptualization, C.-R.L. and K.-P.C.; methodology, C.-R.L. and Y.-W.L.; software, Y.-W.L. and R.G.B.; validation and discussion, C.-R.L., R.G.B., Y.-W.L., I.V.T., J.-H.Y., L.-H.C., D.N.M., P.S.P. and K.-P.C.; investigation, C.-R.L.; writing—original draft preparation, C.-R.L. and K.-P.C.; writing—review and editing, I.V.T., and K.-P.C.; supervision, I.V.T., and K.-P.C.; project administration, I.V.T. and K.-P.C.; funding acquisition, I.V.T. and K.-P.C. All authors have read and agreed to the published version of the manuscript.

Funding: This work is supported by the National Tsing Hua University, Ministry of Education and the National Science and Technology Council (NSTC 111-2923-E-007-008-MY3; 111-2628-E-007-021; 112-2223-E-007-007-MY3; 112-2923-E-007-004-MY2; 112-2119-M-A49-008; 113-2628-E-007-024). This research was funded by the Russian Science Foundation (project no. 22-42-08003).

Data Availability Statement: The data presented in this study are available on request from the corresponding author.

Acknowledgments: Thanks to Wei Lee's support with computation resources in COMSOL.

Conflicts of Interest: The authors declare no conflicts of interest.

References

1. Cui, Y.; Kang, L.; Lan, S.; Rodrigues, S.; Cai, W. Giant chiral optical response from a twisted-arc metamaterial. *Nano Lett.* **2014**, *14*, 1021–1025. [CrossRef] [PubMed]
2. Plum, E.; Fedotov, V.; Zheludev, N. Extrinsic electromagnetic chirality in metamaterials. *J. Opt. A Pure Appl. Opt.* **2009**, *11*, 074009. [CrossRef]
3. Decker, M.; Klein, M.; Wegener, M.; Linden, S. Circular dichroism of planar chiral magnetic metamaterials. *Opt. Lett.* **2007**, *32*, 856–858. [CrossRef]
4. Gorkunov, M.; Antonov, A. Rational design of maximum chiral dielectric metasurfaces. In *All-Dielectric Nanophotonics*; Elsevier: Amsterdam, The Netherlands, 2024; pp. 243–286.
5. Wang, R.; Wang, C.; Sun, T.; Hu, X.; Wang, C. Simultaneous broadband and high circular dichroism with two-dimensional all-dielectric chiral metasurface. *Nanophotonics* **2023**, *12*, 2192–8614. [CrossRef]
6. Chen, M.K.; Wu, Y.; Feng, L.; Fan, Q.; Lu, M.; Xu, T.; Tsai, D.P. Principles, Functions, and Applications of Optical Meta-Lens. *Adv. Opt. Mater.* **2021**, *9*, 2001414. [CrossRef]
7. Kuznetsov, A.I.; Brongersma, M.L.; Yao, J.; Chen, M.K.; Levy, U.; Tsai, D.P.; Zheludev, N.I.; Faraon, A.; Arbabi, A.; Yu, N. Roadmap for Optical Metasurfaces. *ACS Photonics* **2024**, *11*, 2330–4022. [CrossRef]
8. Sharma, V.; Crne, M.; Park, J.O.; Srinivasarao, M. Structural origin of circularly polarized iridescence in jeweled beetles. *Science* **2009**, *325*, 449–451. [CrossRef]
9. Brady, P.; Cummings, M. Differential response to circularly polarized light by the jewel scarab beetle Chrysina gloriosa. *Am. Nat.* **2010**, *175*, 614–620. [CrossRef]
10. Yang, Z.-J.; Hu, D.-J.; Gao, F.-H.; Hou, Y.-D. Enhanced chiral response from the Fabry–Perot cavity coupled meta-surfaces. *Chin. Phys. B* **2016**, *25*, 084201. [CrossRef]
11. Tang, B.; Li, Z.; Palacios, E.; Liu, Z.; Butun, S.; Aydin, K. Chiral-selective plasmonic metasurface absorbers operating at visible frequencies. *IEEE Photonics Technol. Lett.* **2017**, *29*, 295–298. [CrossRef]

12. Wang, Z.; Jia, H.; Yao, K.; Cai, W.; Chen, H.; Liu, Y. Circular dichroism metamirrors with near-perfect extinction. *ACS Photonics* **2016**, *3*, 2096–2101. [CrossRef]
13. Jing, L.; Wang, Z.; Yang, Y.; Zheng, B.; Liu, Y.; Chen, H. Chiral metamirrors for broadband spin-selective absorption. *Appl. Phys. Lett.* **2017**, *110*, 231103. [CrossRef]
14. Long, G.; Adamo, G.; Tian, J.; Klein, M.; Krishnamoorthy, H.N.; Feltri, E.; Wang, H.; Soci, C. Perovskite metasurfaces with large superstructural chirality. *Nat. Commun.* **2022**, *13*, 1551. [CrossRef] [PubMed]
15. Zhao, Y.; Belkin, M.; Alù, A. Twisted optical metamaterials for planarized ultrathin broadband circular polarizers. *Nat. Commun.* **2012**, *3*, 870. [CrossRef]
16. Gansel, J.K.; Latzel, M.; Frölich, A.; Kaschke, J.; Thiel, M.; Wegener, M. Tapered gold-helix metamaterials as improved circular polarizers. *Appl. Phys. Lett.* **2012**, *100*, 101109. [CrossRef]
17. Yang, Z.; Zhao, M.; Lu, P.; Lu, Y. Ultrabroadband optical circular polarizers consisting of double-helical nanowire structures. *Opt. Lett.* **2010**, *35*, 2588–2590. [CrossRef]
18. Kaschke, J.; Gansel, J.K.; Wegener, M. On metamaterial circular polarizers based on metal N-helices. *Opt. Express* **2012**, *20*, 26012–26020. [CrossRef]
19. Gansel, J.K.; Thiel, M.; Rill, M.S.; Decker, M.; Bade, K.; Saile, V.; von Freymann, G.; Linden, S.; Wegener, M. Gold helix photonic metamaterial as broadband circular polarizer. *Science* **2009**, *325*, 1513–1515. [CrossRef]
20. Basiri, A.; Chen, X.; Bai, J.; Amrollahi, P.; Carpenter, J.; Holman, Z.; Wang, C.; Yao, Y. Nature-inspired chiral metasurfaces for circular polarization detection and full-Stokes polarimetric measurements. *Light Sci. Appl.* **2019**, *8*, 78. [CrossRef]
21. Wu, R.; Jiang, K.; Jiang, X.; Xu, J.; Yue, Z.; Teng, S. Metasurface-based circular polarizer with a controllable phase and its application in holographic imaging. *Opt. Lett.* **2024**, *49*, 774–777. [CrossRef]
22. Li, W.; Coppens, Z.J.; Besteiro, L.V.; Wang, W.; Govorov, A.O.; Valentine, J. Circularly polarized light detection with hot electrons in chiral plasmonic metamaterials. *Nat. Commun.* **2015**, *6*, 8379. [CrossRef]
23. Wang, W.; Besteiro, L.V.; Liu, T.; Wu, C.; Sun, J.; Yu, P.; Chang, L.; Wang, Z.; Govorov, A.O. Generation of Hot Electrons with Chiral Metamaterial Perfect Absorbers: Giant Optical Chirality for Polarization-Sensitive Photochemistry. *ACS Photonics* **2019**, *6*, 3241–3252. [CrossRef]
24. Chen, Y.; Yang, X.; Gao, J. 3D Janus plasmonic helical nanoapertures for polarization-encrypted data storage. *Light Sci. Appl.* **2019**, *8*, 45. [CrossRef] [PubMed]
25. Wang, J.; Jiang, H.; Fan, S.; Wu, F.; Zhao, W. Information encryption driven by strength-switchable circular dichroism in vanadium dioxide based chiral metasurface. *Phys. B Condens. Matter* **2024**, *682*, 415864. [CrossRef]
26. Zhao, Y.; Askarpour, A.N.; Sun, L.; Shi, J.; Li, X.; Alù, A. Chirality detection of enantiomers using twisted optical metamaterials. *Nat. Commun.* **2017**, *8*, 14180. [CrossRef] [PubMed]
27. Berova, N.; Nakanishi, K.; Woody, R.W.; Woody, R. *Circular Dichroism: Principles and Applications*; John Wiley & Sons: Hoboken, NJ, USA, 2000.
28. Hendry, E.; Carpy, T.; Johnston, J.; Popland, M.; Mikhaylovskiy, R.V.; Lapthorn, A.J.; Kelly, S.M.; Barron, L.D.; Gadegaard, N.; Kadodwala, M.J.N.N. Ultrasensitive detection and characterization of biomolecules using superchiral fields. *Nat. Nanotechnol.* **2010**, *5*, 783. [CrossRef]
29. Govorov, A.O.; Fan, Z.; Hernandez, P.; Slocik, J.M.; Naik, R.R. Theory of circular dichroism of nanomaterials comprising chiral molecules and nanocrystals: Plasmon enhancement, dipole interactions, and dielectric effects. *Nano Lett.* **2010**, *10*, 1374–1382. [CrossRef]
30. Zhang, M. Chiral biosensing using terahertz twisted chiral metamaterial. *Opt. Express* **2022**, *30*, 14651–14660. [CrossRef]
31. Plum, E.; Zheludev, N.I. Chiral mirrors. *Appl. Phys. Lett.* **2015**, *106*, 221901. [CrossRef]
32. Plum, E. Extrinsic chirality: Tunable optically active reflectors and perfect absorbers. *Appl. Phys. Lett.* **2016**, *108*, 241905. [CrossRef]
33. Sievenpiper, D.; Zhang, L.; Broas, R.F.; Alexopolous, N.G.; Yablonovitch, E. High-impedance electromagnetic surfaces with a forbidden frequency band. *IEEE Trans. Microw. Theory Tech.* **1999**, *47*, 2059–2074. [CrossRef]
34. Feresidis, A.P.; Goussetis, G.; Wang, S.; Vardaxoglou, J.C. Artificial magnetic conductor surfaces and their application to low-profile high-gain planar antennas. *IEEE Trans. Antennas Propag.* **2005**, *53*, 209–215. [CrossRef]
35. Zhu, Z.H.; Guo, C.C.; Liu, K.; Ye, W.M.; Yuan, X.D.; Yang, B.; Ma, T. Metallic nanofilm half-wave plate based on magnetic plasmon resonance. *Opt. Lett.* **2012**, *37*, 698–700. [CrossRef] [PubMed]
36. Pors, A.; Nielsen, M.G.; Bozhevolnyi, S.I. Broadband plasmonic half-wave plates in reflection. *Opt. Lett.* **2013**, *38*, 513–515. [CrossRef]
37. Deng, Y.; Cai, Z.; Ding, Y.; Bozhevolnyi, S.I.; Ding, F. Recent progress in metasurface-enabled optical waveplates. *Nanophotonics* **2022**, *11*, 2219–2244. [CrossRef]
38. Lin, M.-Y.; Xu, W.-H.; Bikbaev, R.G.; Yang, J.-H.; Li, C.-R.; Timofeev, I.V.; Lee, W.; Chen, K.-P. Chiral-Selective Tamm Plasmon Polaritons. *Materials* **2021**, *14*, 2788. [CrossRef]
39. Aydin, K.; Ferry, V.E.; Briggs, R.M.; Atwater, H.A. Broadband polarization-independent resonant light absorption using ultrathin plasmonic super absorbers. *Nat. Commun.* **2011**, *2*, 517. [CrossRef]
40. Lin, C.-H.; Chern, R.-L.; Lin, H.-Y. Polarization-independent broad-band nearly perfect absorbers in the visible regime. *Opt. Express* **2011**, *19*, 415–424. [CrossRef]

41. Liu, N.; Mesch, M.; Weiss, T.; Hentschel, M.; Giessen, H. Infrared perfect absorber and its application as plasmonic sensor. *Nano Lett.* **2010**, *10*, 2342–2348. [CrossRef]
42. Chen, H.-T. Interference theory of metamaterial perfect absorbers. *Opt. Express* **2012**, *20*, 7165–7172. [CrossRef]
43. Hao, J.; Zhou, L.; Qiu, M. Nearly total absorption of light and heat generation by plasmonic metamaterials. *Phys. Rev. B* **2011**, *83*, 165107. [CrossRef]
44. Menzel, C.; Helgert, C.; Rockstuhl, C.; Kley, E.B.; Tünnermann, A.; Pertsch, T.; Lederer, F. Asymmetric transmission of linearly polarized light at optical metamaterials. *Phys. Rev. Lett.* **2010**, *104*, 253902. [CrossRef] [PubMed]
45. Bikbaev, R.G.; Maksimov, D.N.; Chen, K.-P.; Timofeev, I.V. Double-Resolved Beam Steering by Metagrating-Based Tamm Plasmon Polariton. *Materials* **2022**, *15*, 6014. [CrossRef] [PubMed]
46. Huang, Y.-W.; Lee, H.W.H.; Sokhoyan, R.; Pala, R.A.; Thyagarajan, K.; Han, S.; Tsai, D.P.; Harry, A. Atwater. Gate-Tunable Conducting Oxide Metasurfaces. *Nano Lett.* **2016**, *16*, 5319–5325. [CrossRef]

Disclaimer/Publisher's Note: The statements, opinions and data contained in all publications are solely those of the individual author(s) and contributor(s) and not of MDPI and/or the editor(s). MDPI and/or the editor(s) disclaim responsibility for any injury to people or property resulting from any ideas, methods, instructions or products referred to in the content.

Article

Exploiting the Bragg Mirror Effect of TiO$_2$ Nanotube Photonic Crystals for Promoting Photoelectrochemical Water Splitting

Ming Meng [1], Hucheng Zhou [1], Jing Yang [1], Liwei Wang [1], Honglei Yuan [1], Yanling Hao [2,*] and Zhixing Gan [3,4,*]

[1] School of Physics and Telecommunication Engineering, Zhoukou Normal University, Zhoukou 466001, China; mengming@zknu.edu.cn (M.M.); 20182013@zknu.edu.cn (H.Z.); yangjing0410@zju.edu.cn (J.Y.); wangliwei@zknu.edu.cn (L.W.); yhl@zknu.edu.cn (H.Y.)
[2] Key Laboratory for Micro-Nano Functional Materials of Qianxinan, Minzu Normal University of Xingyi, Xingyi 562400, China
[3] Center for Future Optoelectronic Functional Materials, Nanjing Normal University, Nanjing 210023, China
[4] School of Computer and Electronic Information/School of Artificial Intelligence, Nanjing Normal University, Nanjing 210023, China
* Correspondence: haoyanling@xynun.edu.cn (Y.H.); zxgan@njnu.edu.cn (Z.G.)

Citation: Meng, M.; Zhou, H.; Yang, J.; Wang, L.; Yuan, H.; Hao, Y.; Gan, Z. Exploiting the Bragg Mirror Effect of TiO$_2$ Nanotube Photonic Crystals for Promoting Photoelectrochemical Water Splitting. *Nanomaterials* 2024, 14, 1695. https://doi.org/10.3390/nano14211695

Academic Editor: Antonino Gulino

Received: 6 September 2024
Revised: 15 October 2024
Accepted: 21 October 2024
Published: 23 October 2024

Copyright: © 2024 by the authors. Licensee MDPI, Basel, Switzerland. This article is an open access article distributed under the terms and conditions of the Creative Commons Attribution (CC BY) license (https://creativecommons.org/licenses/by/4.0/).

Abstract: Exploiting the Bragg mirror effect of photonic crystal photoelectrode is desperately desired for photoelectrochemical water splitting. Herein, a novel TiO$_2$ nanotube photonic crystal bi-layer structure consisting of a top nanotube layer and a bottom nanotube photonic crystal layer is presented. In this architecture, the photonic bandgap of bottom TiO$_2$ nanotube photonic crystals can be precisely adjusted by modulating the anodization parameters. When the photonic bandgap of bottom TiO$_2$ nanotube photonic crystals overlaps with the electronic bandgap of TiO$_2$, the bottom TiO$_2$ nanotube photonic crystal layer will act as a Bragg mirror, leading to the boosted ultraviolet light absorption of the top TiO$_2$ nanotube layer. Benefiting from the promoted UV light absorption, the TiO$_2$ NT-115-NTPC yields a photocurrent density of 1.4 mA/cm^2 at 0.22 V vs. Ag/AgCl with a Faradic efficiency of 100%, nearly two times higher than that of conventional TiO$_2$ nanotube arrays. Furthermore, incident photon-to-current conversion efficiency is also promoted within ultraviolet light region. This research offers an effective strategy for improving the performance of photoelectrochemical water splitting through intensifying the light–matter interaction.

Keywords: TiO$_2$ nanotube; photonic crystals; Bragg mirror effect; photoelectrochemical water splitting

1. Introduction

Photoelectrochemical (PEC) water splitting has been regarded as a desirable avenue to exploit the abundant and sustainable solar energy by directly transforming the incident light into hydrogen fuel [1–7]. Among various PEC materials that can work as photoelectrodes, TiO$_2$ nanotube array (TiO$_2$ NTA), vertically grown on a conductive Ti foil by electrochemical anodization, distinguishes itself owing to its multifold unique merits including large internal and external surface areas, unidirectional electrical channel, and outstanding adhesion with Ti foil [8–10]. Unfortunately, the pristine TiO$_2$ NTA has a wide bandgap of 3.2 eV, which means that it still suffers from limited photoconversion efficiency even in the ultraviolet (UV) region [11–13]. Until now, element doping and narrow bandgap semiconductor coupling have been the two major strategies for extending the visible light absorption of TiO$_2$ [14–17]. However, these strategies bear many adverse effects, such as limited visible light response, increased carrier recombination centers, decreased incident photon-to-electron conversion efficiency (*IPCE*) in the UV region and the redox ability of the photogenerated charges [18–21]. Accordingly, gaining the utmost out of the UV light may be another promising approach to promote the PEC performance of TiO$_2$ NTA.

Recently, introducing a photonic crystal nanostructure into photocatalysts furnishes a new emerging route of strengthening light–matter interaction to improve light absorption [22–24]. The photonic crystal photocatalysts possess a periodic dielectric structure, which endows them with a photonic bandgap (PBG) for a certain frequency of photons [25–27]. To be specific, the group velocity of the photons with the frequency near the PBG edges can be significantly slowed, referred as the slow photon effect [28–31]. In addition, the photons with the frequency range of PBG are totally reflected and cannot propagate in the photonic crystal structure due to Bragg reflection (called the Bragg mirror effect) [32–34]. Obviously, the slow photon effect and Bragg mirror effect hold immense promise for intensifying light–material interaction, resulting in an amplified light absorption and photoelectrochemical reaction. Yet, to date, the existing investigations have mainly focused on the utilization of the slow photon effect in a single layer of three-dimensional (3D) TiO_2 inverse opal structures [13,28,35–37], which cannot use the reflected light at the PBG of TiO_2 to promote PEC performance in the UV region.

Apart from the 3D TiO_2 inverse opal structures, novel TiO_2 nanotube photonic crystals (TiO_2 NTPCs) with periodicities along the axial direction of nanotube have been successfully fabricated by a simple periodic current pulse anodization process [32,38,39]. Furthermore, the PBGs of TiO_2 NTPCs can be continuously adjusted through controlling the fabrication parameters [32,38,39]. Undoubtedly, after constructing the TiO_2 NTPC bi-layer structure consisting of a top nanotube (NT), which functions as an absorbing layer, and the bottom NTPC with PBG overlapping with an electronic bandgap of TiO_2 that acts as Bragg mirror layer, the interaction of top TiO_2 NT layer with reflected UV light should be greatly boosted, which could enhance its PEC performance in the UV region. Nevertheless, there is as yet no investigation available on the TiO_2 NTPC bi-layer structure focusing on the correlation between the Bragg mirror effect and PEC performance. This also implies that the underlying physical mechanism also remains unclear.

Herein, the novel TiO_2 NTPC bi-layer structure consisting of a top NT layer and a bottom NTPC layer was designed and fabricated for PEC water splitting. As expected, the TiO_2 NTPC bi-layer structure, with the PBG of bottom NTPC overlapping with an electronic bandgap of TiO_2, yielded a photocurrent density of 1.4 mA/cm^2 at 0.22 V vs. Ag/AgCl with Faradic efficiency of 100%, nearly two times higher than that of conventional TiO_2 NTA. Furthermore, *IPCE* was also promoted within the UV light region. Such remarkable enhancement of PEC water splitting activity was primarily derived from the fact that the bottom NTPC layer can function as a Bragg mirror that can promote the interaction of top TiO_2 NT layer with the reflected UV light, thus leading to the boosted UV light absorption of the top TiO_2 NT layer. This work offers an effective strategy for improving the performance of PEC water splitting through intensifying light–matter interaction.

2. Materials and Methods

2.1. Materials

Ammonium fluoride (NH_4F), Ethylene glycol and Sodium hydroxide (NaOH) were obtained from Sinopharm Chemical Reagent Co., Ltd., Shanghai, China. Ti foils (0.25 mm thick, 99.8% purity) were purchased from Anping Anheng Wire Mesh Co., Ltd., Hengshui, China. All the chemicals were utilized as received without any further purification.

2.2. Fabrication of the TiO_2 NTPC Bi-Layer Structure

The TiO_2 NTPC bi-layer structure consisting of a top nanotube (NT) layer and a bottom NTPC layer was prepared by successive two-step anodization. Specifically, the Ti foils with sizes of 1.5 cm × 1 cm were firstly pre-treated by anodization at 60 V for 1 h, using glycol aqueous solution containing 0.5 wt% NH_4F and 2 vol% DI H_2O. Then, the as-grown TiO_2 NT was ultrasonically removed in deionized (DI) H_2O. After that, the pre-treated Ti foils were subjected to a successive two-step anodization process composed of a constant current anodization part and subsequently a periodic current pulse anodization part. During the first step, the constant current was maintained at 7 mA/cm^2 for 10 min to form the top TiO_2

NT layer. During the second step, the periodic current pulse anodization with high current (HC, $J_{HC} = 7$ mA/cm^2) and low current (LC, $J_{HC} = 0$ mA/cm^2) was employed to fabricate the bottom TiO$_2$ NTPC layer. The time duration of the HC pulse was controlled from 120 to 180 s, while the time duration of the LC pulse was fixed at 180 s to tailor the lattice constant of NTPC. Finally, the bi-layer structures were annealed at 450 °C in air for 2 h to obtain anatase TiO$_2$. In addition, the single layer TiO$_2$ NTPC with the same thickness as the bottom NTPC layer of the bi-layer was also prepared by constant current anodization.

2.3. Characterization

The morphologies, microstructures and crystal structures of the as-prepared samples were inspected by field-emission scanning electron microscopy (FE-SEM, S4800, Hitachi Ltd., Tokyo, Japan), field-emission transmission electron microscopy (FE-TEM, JEM-2100, JEOL Ltd., Tokyo, Japan), and X-ray powder diffractometry (XRD, Xpert, Philips, Amsterdam, The Netherlands). The diffuse reflectance spectra were recorded by a VARIAN Cary5000 spectrophotometer (Varian, CA, USA). The X-ray photoelectron spectroscopy (XPS) data were collected by a PHI 5000 Versaprobe (Ulvac-Phi, Kanagawa, Japan).

2.4. Photoelectrochemical Measurements

Photoelectrochemical measurements were performed in a three-electrode system connected to a CHI 660E electrochemical workstation (CH Instrument, Chenhua Ltd., Shanghai, China) utilizing the as-prepared samples with an exposed area of 1 cm^2 as the working electrode, the Pt mesh as the counter electrode, and the Ag/AgCl (3 mol/L KCl-filled) as the reference electrode. The 1 M NaOH (pH = 13.6) solution was electrolyte, which was purged with N$_2$ (99.999%) flow for 1 h to remove dissolved oxygen. The illumination source was a 500 W Xe lamp (Solar 500, NBet Group Corp. Beijing, China) with a calibrated intensity of 100 mW/cm^2, and a water filter was placed between the lamp and the electrochemical cell to eliminate the infrared heating of the electrolyte. The incident photon-to-current conversion efficiency ($IPCE$) measurements were conducted at an applied potential of 0.22 V vs. Ag/AgCl by means of a monochromatic system. During the PEC stability measurement, the photoelectrodes were biased at 0.22 V vs. Ag/AgCl. The amount of evolved oxygen was quantified by an Ocean Optics oxygen sensor system equipped with a FOXY probe (NeoFox Phase Measurement System), which was measured together with PEC stability.

3. Results and Discussion

3.1. Morphological Characterization of the TiO$_2$ NTPC

Figure 1a,b are a schematic illustration of the current–time curve of anodization and TiO$_2$ nanotube photonic crystal (NTPC) bi-layer structure consisting of a top NT and a bottom NTPC. In brief, the single constant current anodization was first utilized to form the top smooth-walled TiO$_2$ NT layer on the Ti foil. The as-grown sample was then subjected to periodic current pule anodization with high current and low current to form a TiO$_2$ NTPC layer with periodicities along the axial direction beneath the TiO$_2$ NT layer. The most crucial step in this work was to accurately modulate the lattice constant of the TiO$_2$ NTPC for obtaining the desired featured. This could be realized by adjusting the duration of high current pulse anodization, since the lattice constant of the NTPC layer was almost linearly increased with it.

Figure 1c–e display the FE-SEM images of single-layer TiO$_2$ NTPC fabricated by the periodic current pulse anodization with different durations of the HC pulse of 7 mA/cm^{-2}. Obviously, a TiO$_2$ NTPC presents a bamboo-shaped periodic structure in the axial direction of the NT. Such a periodic structure with alternating protrusive bamboo node layers and smooth-walled tube layers can result in a periodical refractive index change in the longitudinal direction, indicating that it can exhibit a structural modulated photonic bandgap (PBG). The length of the node and smooth-walled layer is the lattice constant of TiO$_2$ NTPC, which increases from 115 nm to 180 nm for HC pulse durations of 120 s and 180 s, respectively. The corresponding samples are denoted as 115-NTPC and 180-NTPC, respectively. Additionally, these TiO$_2$ NTPC have well-ordered and hexagonally arranged tubular structures with an average diameter of about 100 nm and a wall thickness of about 10 nm. More importantly, this allows the PBG of TiO$_2$ NTPC to be adjusted at will to match the electronic bandgap of anatase TiO$_2$.

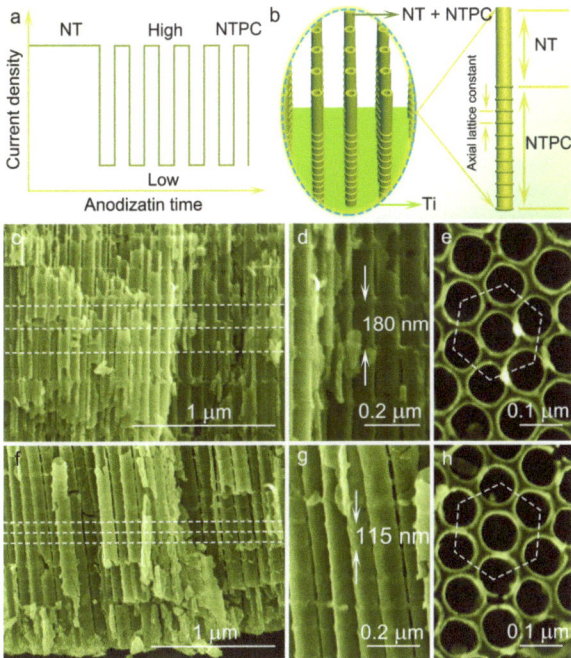

Figure 1. (a) A schematic illustration of the current–time curve of anodization and the TiO$_2$ nanotube/nanotube photonic crystal bi-layer structure. (b) Schematic illustration of the TiO$_2$ nanotube/nanotube photonic crystal bi-layer structure. (c–e) FE-SEM images of the TiO$_2$ 180-NTPC. (f–h) FE-SEM images of the TiO$_2$ 115-NTPC.

3.2. Optical Obsorption Properties of the TiO$_2$ NTPC Structure

The reflectance spectra of TiO$_2$ NTPCs with different lattice constants in air, ethanol and electrolytes are measured under normal incidence and are presented in Figure 2a. As the lattice constant increases from 115 nm to 180 nm, the refection peak of the TiO$_2$ NTPC shifts to longer wavelengths. The positions of PBG of 115-NTPC with 15 periods and 180-NTPC with 15 periods are located at around 378 and 462 nm, respectively. The reflectance spectra of TiO$_2$ NTPC are strongly influenced by the refractive index contrast [40,41]. When the TiO$_2$ NTPC was put in ethanol, a remarkable red-shift of the reflection peaks could be observed, compared with that sample in air. The result is further reflected from the colors of the TiO$_2$ NTPCs (Figures 2b,c and S1). Specifically, after being infiltrated with ethanol, its color changes from purple to green. It should be noted that no color change in TiO$_2$ 115-NTPC can be found, since its PBG is in the ultraviolet region (below 400 nm). When the refractive index contrast is further reduced by infiltration with liquid electrolyte (1 M NaOH), the reflection peaks of TiO$_2$ NTPC shift to an even longer wavelength. Taking TiO$_2$ 115-NTPC as an example, when the sample is immersed in electrolyte, the position of PBG shifts from 378 nm (air) to 384 (1 M NaOH) nm, which is very close to the electronic bandgap of TiO$_2$.

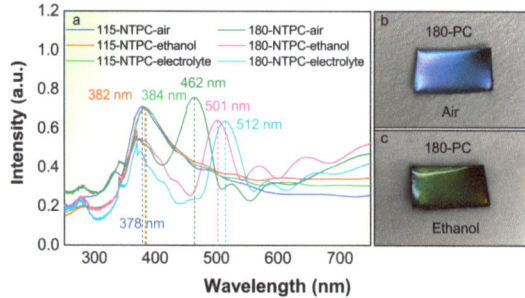

Figure 2. (a) The reflectance spectra of the TiO$_2$ 180-NTPC and TiO$_2$ 115-NTPC samples in air and infiltrated with ethanol and electrolytes, respectively. (b,c) Photographs of the TiO$_2$ 180-NTPC samples in air and infiltrated with ethanol, respectively.

3.3. Microstructure, Crystalline Phase, and Chemical Composition Analysis of the TiO$_2$ NTPC

The microstructure, crystalline phase, and chemical composition of the TiO$_2$ NTPCs are also analyzed by FE-TEM, XRD, and XPS. The low-magnification FE-TEM images further confirm that the TiO$_2$ NTPC samples show a hexagonally arranged and bamboo-shaped periodic structure in the axial direction, which is consistent with FE-SEM results (Figures 3a,b and S2). The HR-TEM image reveals that the well-resolved lattice spacing of 0.35 nm matches the d-spacing of the (101) plane of the anatase TiO$_2$, which is further proved by the corresponding Fast-Fourier Transform diffraction pattern (inset of Figure 3b) [42,43]. Figure 3c presents the XRD pattern of the TiO$_2$ NTPC, suggesting that all the diffraction peaks can be indexed to the anatase TiO$_2$ (JCPDS 21-1276) except those from the Ti substrate [44–46]. The XPS spectra also demonstrate that the TiO$_2$ NTPC samples are pure anatase with some oxygen deficiencies (Figure S3) [47].

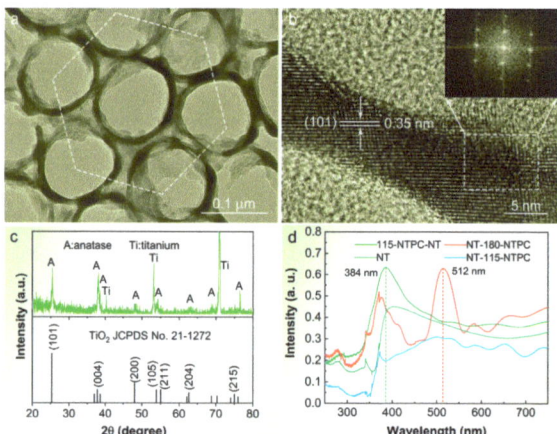

Figure 3. (a) A low-magnification FE-TEM image of the TiO$_2$ 115 NTPC. (b) A HR-TEM image of the area highlighted by the white dashed hexagon in (a). Inset: Fast-Fourier Transform diffraction patterns of the areas bounded by the white dashed box in (b). (c) Corresponding XRD pattern. (d) Reflectance spectra of the TiO$_2$ NT-115-NTPC, TiO$_2$ NT-180-NTPC, TiO$_2$ 115-NTPC-NT, and TiO$_2$ NT infiltrated with electrolytes, respectively.

3.4. Morphological Characterizaiton and Optical Obsorption Properties of the TiO$_2$ NTPC Bi-Layer Structure

To confirm that the PBG reflection effect can lead to a significant enhancement in light absorption, we further fabricated the TiO$_2$ NTPC bi-layer structure by successive two-step anodization. Figure S4 depicts the cross-sectional FE-SEM image of the TiO$_2$ 115-NTPC bi-layer structure (referred as TiO$_2$ NT-115-NTPC). The TiO$_2$ NT-115-NTPCs with 15 periods can be clearly seen to seamlessly grow beneath the smooth-walled NT layer with a thickness of approximately 500 nm, ensuring excellent connection between the two layers and easy electrolyte infiltration. For comparison, we also fabricated a TiO$_2$ 180-NTPC bi-layer structure (referred as TiO$_2$ NT-180-NTPC), another TiO$_2$ 115-NTPC bi-layer structure consisting of top NTPC layer and bottom NTs (TiO$_2$ 115-NTPC-NT), and a TiO$_2$ NT without a photonic crystal layer. To gain more realistic insights into the optical properties of TiO$_2$ NT-115-NTPC, TiO$_2$ NT-180-NTPC, TiO$_2$ 115-NTPC-NT, and TiO$_2$ NT, the reflectance spectra of the four samples in electrolytes were examined. As shown in Figure 3d, the TiO$_2$ NT-115-NTPC exhibits the strongest UV light harvesting capacity among the aforementioned four samples. This could be mainly attributed to the Bragg mirror effect of the bottom TiO$_2$ 115-NTPC with the PBG (3.2 eV) coinciding with the electronic bandgap of anatase TiO$_2$ (3.2 eV) that can reflect the UV light back to the absorbing NT layer, thus leading to the boosted UV light absorption of the top TiO$_2$ NT layer. Enhanced light absorption has been found in the TiO$_2$ NTPC bi-layer structure-based dye-sensitized solar cells [48].

3.5. PEC Water Splitting Activity of the TiO$_2$ NTPC Structure

To determine the promoted PEC performance of TiO$_2$ NT-115-NTPC, a set of PEC measurements were carried out in a three-electrode configuration using the as-prepared samples, Pt mesh, and Ag/AgCl (3 mol L^{-1} KCl-filled) as the working, counter, and reference electrodes, respectively. The electrolytes for the PEC water splitting reaction were an aqueous solution of 1M NaOH (pH = 13.6). Figure 4a displays the linear-sweep voltammogram (LSV) sweeps for TiO$_2$ NT-115-NTPC, TiO$_2$ NT-180-NTPC, TiO$_2$ 115-NTPC, and TiO$_2$ NT under light irradiation and dark conditions. All the samples produced almost negligible dark current in comparison with their photocurrent, suggesting no occurrence of electrocatalytic water splitting. Under irradiation, the photocurrent density of TiO$_2$ NT-115-NTPC sharply increased and largely surpassed those of TiO$_2$ NT-180-NTPC, TiO$_2$ 115-NTPC, and TiO$_2$ NT, which signifies that the TiO$_2$ NT-115-NTPC had the highest PEC performance among the four samples. To elucidate this phenomenon more distinctly, their transient photocurrent responses were also measured under illumination with several 10 s light on/off cycles at 0.22 V vs. Ag/AgCl [1.23 V vs. RHE (reversible hydrogen electrode)], and the results are presented in Figure 4b.

At 0.22 V vs. Ag/AgCl, the TiO$_2$ NT-115-NTPC delivered a maximal photocurrent density of 1.4 mA/cm^2, and it was about 2.05, 2.15 and 3.5 times those of TiO$_2$ NT TiO$_2$ NT-180-NTPC and TiO$_2$ 115-NTPC, respectively. The low photocurrent of the TiO$_2$ NT-180-NTPC can be ascribed to the fact that its PBG position of the bottom TiO$_2$ 180-NTPC was outside of the electronic bandgap of anatase TiO$_2$ (3.2 eV), meaning that it could not reflect the UV light back to the top NT layer. That is to say, the bottom TiO$_2$ 180-NTPC had no impact on the PEC performance of TiO$_2$ NT-180-NTPC. Compared with TiO$_2$ NT, the low photocurrent of the TiO$_2$ 115-NTPC was mainly due to the single NTPC layer with a PGB position of 384 nm, which resulted in the decrease in UV light absorption.

To visualize the photocurrent enhancement owing to the promoted UV light absorption, incident photon-to-current conversion efficiency (*IPCE*) measurements were conducted on TiO$_2$ NT-115-NTPC and TiO$_2$ NT at 0.22 V vs. Ag/AgCl. The *IPCE* could be calculated as a percentage according to the following equation [49]:

$$IPCE = \frac{1240I}{\lambda J_{light}} \quad (1)$$

where *I* is the measured photocurrent density at a specific wavelength, λ is the wavelength of the incident light, and J_{light} is the light intensity of a specific wavelength. As shown in Figure 4c, the TiO$_2$ NT-115-NTPC has greatly boosted *IPCE* values only in the UV region and reaches its highest value of 96% at 380 nm compared with TiO$_2$ NT. The result provides a clue suggesting that the bottom TiO$_2$ 115-NTPC makes a great contribution to PEC performance in the UV light region.

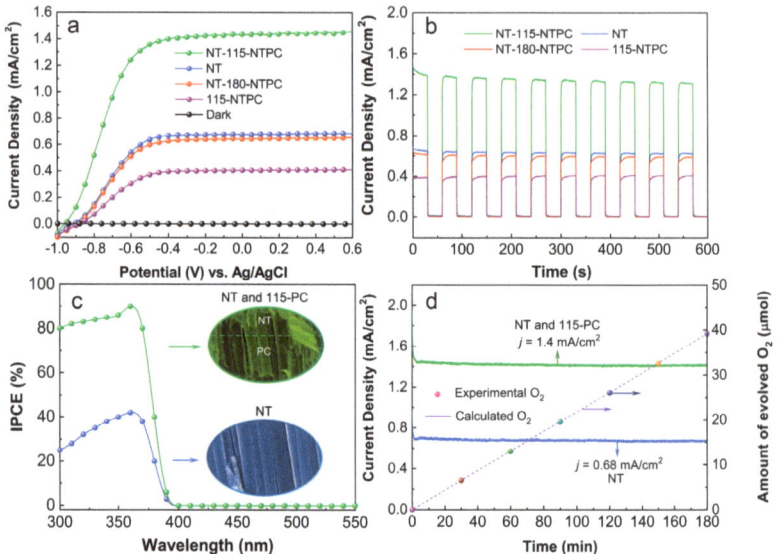

Figure 4. (a) Current vs. voltage (J−V) curves acquired from TiO$_2$ NT-115-NTPC, TiO$_2$ NT-180-NTPC, TiO$_2$ 115-NTPC, TiO$_2$ NT, respectively. (b) The corresponding transient photocurrent responses performed at 0.22 vs. Ag/AgCl. (c) The IPCE spectra of TiO$_2$ NT-115-NTPC and TiO$_2$ NT measured at an incident wavelength range from 300 nm to 550 nm at a potential 0.22 V vs. Ag/AgCl. (d) Photocurrent versus time (J−t) curves of TiO$_2$ NT-115-NTPC and TiO$_2$ NT obtained at 0.22 V vs. RHE. The dashed line and colorful spheres show the amount of evolved O$_2$ calculated theoretically and detected experimentally for TiO$_2$ NT-115-NTPC, respectively.

PEC stability and Faradic efficiency are two important parameters for the practical application of photoelectrode. Figure 4d presents the photocurrent–time curves of TiO$_2$ NT-115-NTPC and TiO$_2$ NT measured at 0.22 V vs. Ag/AgCl and continuous light illumination. The photocurrent densities of both samples are very stable, and there is no indication of deterioration during the entirely measured 3 h. To clarify whether the observed photocurrent originates from the oxygen evolution reaction, the fluorescence sensor is employed to determine the amount of oxygen evolved from the TiO$_2$ NT-115-NTPC. The amount of evolved oxygen increases linearly with the illumination time with unity Faradic efficiency. In addition, the surface morphology and crystal phase of the TiO$_2$ NT-115-NTPC remain intact after PEC water splitting for 3 h (Figure S5), illustrating that the TiO2 NT-115-NTPC possess prominent stability in the oxygen evolution reaction.

3.6. PEC Water Splitting Activity Mechanism

Based on the above experimental results, the promoted PEC performance of TiO$_2$ NT-115-NTPC can be mainly attributed to the significant enhancement of UV light absorption induced by the its bi-layer structure consisting of a top NT layer and a bottom NTPC layer (Figure 5a). As shown in Figure S6, the PBG (3.2 eV) of the bottom TiO$_2$ 115-NTPC overlaps with the electronic bandgap of anatase TiO$_2$ (3.2 eV). When the UV light strikes the TiO$_2$ NT-115-NTPC, a portion of UV light is absorbed by the top TiO$_2$ NT, producing the photoexcited electrons and holes, whereas another portion of UV light penetrates the top TiO$_2$ NT layer and is reflected by the bottom the 115-NTPC layer serving as the Bragg mirror. In such cases, the reflected light can be absorbed again by the top TiO$_2$ NT, hence promoting UV light absorption by the top TiO$_2$ NT. Additionally, optical interference occurs when UV light is being transmitted and reflected, which leads to strong UV photon resonance modes in the top NT absorbing layer, thus also boosting UV light absorption by the top TiO_2 NT (Figure 5b,c). Similar phenomena have been confirmed for other opal photonic crystal photocatalysis [50–52]. Accordingly, more photoexcited electrons and holes are generated, and remarkable promotion of PEC performance is achieved for the TiO_2 NT-115-NTPC.

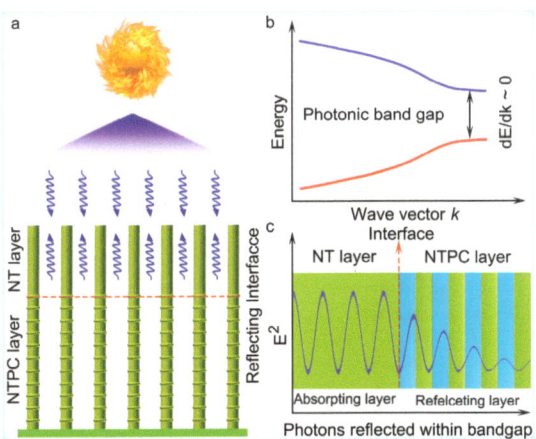

Figure 5. (a) A schematic drawing of mechanism of boosted UV light absorption of TiO_2 NT-115 NTPC. Down arrow and up arrow represent the incident light and reflected light, respectively. (b) Schematic optical band structure of TiO_2 nanotube photonic crystal. (c) Photons reflected within bandgap for further absorption.

4. Conclusions

In summary, we designed and fabricated a novel TiO_2 NTPC bi-layer structure photoanode consisting of a top NT layer and a bottom NTPC layer. In this architecture, when the PBG of bottom NTPC overlapped the with electronic bandgap of TiO_2, the bottom TiO_2 NTPC produced the Bragg mirror effect, leading to boosted UV light harvesting of top TiO_2 NT layer. Benefiting from promoted UV light absorption, the TiO_2 NT-115-NTPC yielded a photocurrent density of 1.4 mA/cm^2 at 0.22 V vs. Ag/AgCl with a Faradic efficiency of 100%, nearly two times higher than that of conventional TiO_2 NT. Furthermore, *IPCE* was also promoted within UV light region. This work provides an effective strategy for improving PEC water splitting through intensifying light–matter interaction.

Supplementary Materials: The following supporting information can be downloaded at https://www.mdpi.com/article/10.3390/nano14211695/s1, Figure S1: Photographs of the TiO_2 115-NTPC in air and infiltrated with ethanol, respectively. Figure S2: (a) A low-magnification FE-TEM image of the TiO_2 180-NTPC. (b) An HR-TEM image of the area highlighted by the white dashed box in (a). (c) Fast-Fourier Transform diffraction patterns of the areas bounded by the white dashed box in (b). (d) Projected atomic models in [101] directions. Figure S3: (a) The XPS survey spectra of the TiO_2 NT-115-NTPC. (b) The corresponding normalized Ti 2p XPS spectra and normal O 1s XPS spectra. Figure S4: FE-SEM images of the TiO_2 NT-115-NTPC. Figure S5: (a) A FE-SEM image of the TiO_2 NT-115-NTPC after undergoing the PEC water splitting reaction for 180 min. (b,c) The corresponding FE-TEM image.

Author Contributions: Conceptualization, M.M., Y.H. and Z.G.; methodology, H.Z. and J.Y.; investigation, L.W.; writing-original draft preparation, H.Y. All authors have read and agreed to the published version of the manuscript.

Funding: This work was supported by National Natural Science Foundation of China (12305006), the Program for Science and Technology Innovation Talents in Universities of Henan Province (24HASTIT012), the Graduate Education Reform Project of Henan Province (2023SJGLX077Y), and the Talent Project of the Guizhou Provincial Education Department (2022-094).

Data Availability Statement: Data are available from the authors on request.

Conflicts of Interest: The authors declare no conflicts of interests.

References

1. Zeng, G.S.; Pham, T.A.; Vanka, S.; Liu, G.J.; Song, C.Y.; Cooper, J.K.; Mi, Z.; Ogitsu, T.; Toma, F.M. Development of a photoelectrochemically self-improving Si/GaN photocathode for efficient and durable H_2 production. *Nat. Mater.* **2021**, *20*, 1130–1135. [CrossRef] [PubMed]
2. Zhao, S.X.; Liu, B.; Li, K.L.; Wang, S.J.; Zhang, G.; Zhao, Z.J.; Wang, T.; Gong, J.L. A silicon photoanode protected with TiO_2/stainless steel bilayer stack for solar seawater splitting. *Nat. Commun.* **2024**, *15*, 2970. [CrossRef] [PubMed]
3. Li, T.T.; Cui, J.Y.; Xu, M.X.; Song, K.P.; Yin, Z.H.; Meng, C.; Liu, H.; Wang, J.J. Efficient acidic photoelectrochemical water splitting enabled by Ru single atoms anchored on hematite photoanodes. *Nano Lett.* **2024**, *24*, 958–965. [CrossRef] [PubMed]
4. Daemi, S.; Kaushik, S.; Das, S.; Hamann, T.W.; Osterloh, F.E. $BiVO_4$-liquid junction photovoltaic cell with 0.2% solar energy conversion efficiency. *J. Am. Chem. Soc.* **2023**, *145*, 25797–25805. [CrossRef]
5. Wang, Y.J.; Wu, Y.P.; Schwartz, J.; Sung, S.H.; Hovden, R.; Mi, Z. A single-junction cathodic approach for stable unassisted solar water splitting. *Joule* **2019**, *3*, 2444–2456. [CrossRef]
6. Meng, M.; Wang, L.W.; Li, C.Y.; Xu, K.; Chen, Y.Y.; Li, J.T.; Gan, Z.X.; Yuan, H.L.; Liu, L.Z.; Li, J. Boosting charge separation on epitaxial In_2O_3 octahedron-nanowire crystal facet-based homojunctions for robust photoelectrochemical water splitting. *Appl. Catal. B Environ.* **2023**, *321*, 120717. [CrossRef]
7. Li, Y.Y.; Zhou, H.; Cai, S.H.; Prabhakaran, D.; Niu, W.T.; Large, A.; Held, G.; Taylor, R.A.; Wu, X.P.; Tsang, S.C.E. Electrolyte-assisted polarization leading to enhanced charge separation and solar-to-hydrogen conversion efficiency of seawater splitting. *Nat. Catal.* **2024**, *7*, 77–88. [CrossRef]
8. Zhao, Y.; Hoivik, N.; Wang, K.Y. Recent advance on engineering titanium dioxide nanotubes for photochemical and photoelectrochemical water splitting. *Nano Energy* **2016**, *30*, 728–744. [CrossRef]
9. Hou, X.L.; Li, Z.; Fan, L.J.; Yuan, J.S.; Lund, P.D.; Li, Y.D. Effect of Ti foil size on the micro sizes of anodic TiO_2 nanotube array and photoelectrochemical water splitting performance. *Chem. Eng. J.* **2021**, *425*, 131415. [CrossRef]
10. Li, F.; Dong, B.; Yu, L.T.; Jin, X.L.; Huang, Q.Z. Construction of photo-thermo-electro coupling field based on surface modification of hydrogenated TiO_2 nanotube array photoanode and its improved photoelectrochemical water splitting. *Inorg. Chem.* **2024**, *63*, 1175–1187. [CrossRef]
11. Hwang, Y.J.; Hahn, C.; Liu, B.; Yang, P.D. Photoelectrochemical properties of TiO_2 nanowire arrays: A study of the dependence on length and atomic layer deposition coating. *ACS Nano* **2016**, *6*, 5060–5069. [CrossRef] [PubMed]
12. Tang, R.; Zhou, S.J.; Zhang, L.Y.; Yin, L.W. Metal-organic framework derived narrow bandgap cobalt carbide sensitized titanium dioxide nanocage for superior photo-electrochemical water oxidation performance. *Adv. Funct. Mater.* **2018**, *28*, 1706154. [CrossRef]
13. Zhang, X.; Liu, Y.; Lee, S.T.; Yang, S.H.; Kang, Z.H. Coupling surface plasmon resonance of gold nanoparticles with slow-photon-effect of TiO_2 photonic crystals for synergistically enhanced photoelectrochemical water splitting. *Energy Environ. Sci.* **2014**, *7*, 1409–1419. [CrossRef]
14. Zhang, W.; He, H.L.; Li, H.Z.; Duan, L.L.; Zu, L.H.; Zhai, Y.P.; Li, W.; Wang, L.Z.; Fu, H.G.; Zhao, D.Y. Visible-light responsive TiO_2-based materials for efficient solar energy utilization. *Adv. Energy Mater.* **2021**, *11*, 2003303. [CrossRef]
15. Zhang, X.F.; Wang, Y.N.; Liu, B.S.; Sang, Y.H.; Liu, H. Heterostructures construction on TiO_2 nanobelts: A powerful tool for building high-performance photocatalysts. *Appl. Catal. B Environ.* **2017**, *202*, 620–641. [CrossRef]
16. Lin, S.W.; Tong, M.H.; Chen, Y.X.; Chen, R.; Zhao, H.P.; Jiang, X.; Yang, K.; Lu, C.Z. CeO_2/TiO_2 heterojunction nanotube arrays for highly efficient visible-light photoelectrochemical water splitting. *ACS Appl. Energy Mater.* **2023**, *6*, 1093–1102. [CrossRef]
17. Park, J.S.; Lee, T.H.; Kim, C.; Lee, S.A.; Choi, M.J.; Kim, H.; Yang, J.W.; Lim, J.; Jang, H.W. Hydrothermally obtained type-II heterojunction nanostructures of In_2S_3/TiO_2 for remarkably enhanced photoelectrochemical water splitting. *Appl. Catal. B Environ.* **2021**, *295*, 120276. [CrossRef]
18. Kang, Q.; Cao, J.Y.; Zhang, Y.J.; Liu, L.Q.; Xu, H.; Ye, J.H. Reduced TiO_2 nanotube arrays for photoelectrochemical water splitting. *J. Mater. Chem. A* **2013**, *1*, 5766–5774. [CrossRef]
19. Sun, B.J.; Zhou, W.; Li, H.Z.; Ren, L.P.; Qiao, P.Z.; Li, W.; Fu, H.G. Synthesis of particulate hierarchical tandem heterojunctions toward optimized photocatalytic hydrogen production. *Adv. Mater.* **2018**, *30*, 1804282. [CrossRef]
20. Lin, T.Q.; Yang, C.Y.; Wang, Z.; Yin, H.; Lu, X.J.; Huang, F.Q.; Lin, J.H.; Xie, X.M.; Jiang, M.H. Effective nonmetal incorporation in black titania with enhanced solar energy utilization. *Energy Environ. Sci.* **2014**, *7*, 967–972. [CrossRef]
21. Li, W.B.; Feng, C.; Dai, S.Y.; Yue, J.G.; Hua, F.X.; Hou, H. Fabrication of sulfur-doped g-C_3N_4/Au/CdS Z-scheme photocatalyst to improve the photocatalytic performance under visible light. *Appl. Catal. B Environ.* **2015**, *168-169*, 465–471. [CrossRef]

22. He, X.H.; Tian, W.; Yang, L.; Bai, Z.Y.; Li, L. Optical and electrical modulation strategies of photoelectrodes for photoelectrochemical water splitting. *Small Methods* **2024**, *8*, 2300350. [CrossRef] [PubMed]
23. Madanu, T.L.; Mouchet, S.R.; Deparis, O.; Liu, J.; Li, Y.; Su, B.L. Tuning and transferring slow photons from TiO_2 photonic crystals to $BiVO_4$ nanoparticles for unprecedented visible light photocatalysis. *J. Colloid Interface Sci.* **2023**, *634*, 290–299. [CrossRef] [PubMed]
24. Likodimos, V. Photonic crystal-assisted visible light activated TiO_2 photocatalysis. *Appl. Catal. B Environ.* **2018**, *230*, 269–303. [CrossRef]
25. Zhang, Z.H.; Zhang, L.B.; Hedhili, M.N.; Zhang, H.N.; Wang, P. Plasmonic gold nanocrystals coupled with photonic crystal seamlessly on TiO_2 nanotube photoelectrodes for efficient visible light photoelectrochemical water splitting. *Nano Lett.* **2013**, *13*, 14–20. [CrossRef]
26. Liu, L.; Lim, S.Y.; Law, C.S.; Jin, B.; Abell, A.D.; Ni, G.; Santos, A. Engineering of broadband nanoporous semiconductor photonic crystals for visible-light-driven photocatalysis. *ACS Appl. Mater. Interfaces* **2020**, *12*, 57079–57092. [CrossRef]
27. Liu, L.; Lim, S.Y.; Law, C.S.; Jin, B.; Abell, A.D.; Ni, G.; Santos, A. Light-confining semiconductor nanoporous anodic alumina optical microcavities for photocatalysis. *J. Mater. Chem. A* **2019**, *7*, 22514–22529. [CrossRef]
28. Chen, X.Q.; Ye, J.H.; Ouyang, S.X.; Kako, T.; Li, Z.S.; Zou, Z.G. Enhanced incident photon-to-electron conversion efficiency of tungsten trioxide photoanodes based on 3D-photonic crystal design. *ACS Nano* **2011**, *5*, 4310–4318. [CrossRef]
29. Zhao, H.; Hu, Z.Y.; Liu, J.; Li, Y.; Wu, M.; Tendeloo, G.V.; Su, B.L. Blue-edge slow photons promoting visible-light hydrogen production on gradient ternary 3DOM TiO_2-Au-CdS photonic crystals. *Nano Energy* **2018**, *47*, 266–274. [CrossRef]
30. Chen, J.I.L.; Von Freymann, G.; Choi, S.Y.; Kitaev, V.; Ozin, G.A. Amplified photochemistry with slow photons. *Adv. Mater.* **2006**, *18*, 1915–1919. [CrossRef]
31. Nan, F.; Kang, Z.H.; Wang, J.L.; Shen, M.R.; Fang, L. Carbon quantum dots coated $BiVO_4$ inverse opals for enhanced photoelectrochemical hydrogen generation. *Appl. Phys. Lett.* **2015**, *106*, 153901. [CrossRef]
32. Chiarello, G.L.; Zuliani, A.; Ceresoli, D.; Martinazzo, R.; Selli, E. Exploiting the photonic crystal properties of TiO_2 nanotube arrays to enhance photocatalytic hydrogen production. *ACS Catal.* **2016**, *6*, 1345–1353. [CrossRef]
33. Chen, Y.W.; Li, L.L.; Xu, Q.L.; Chen, W.; Dong, Y.Q.; Fan, J.J.; Ma, D.K. Recent advances in opal/inverted opal photonic crystal photocatalysts. *Solar PRL* **2021**, *5*, 2000541. [CrossRef]
34. Madanu, T.L.; Chaabane, L.; Mouchet, S.R.; Deparis, O.; Su, B.L. Manipulating multi-spectral slow photons in bilayer inverse opal TiO_2@$BiVO_4$ composites for highly enhanced visible light photocatalysis. *J. Colloid Interface Sci.* **2023**, *647*, 233–245. [CrossRef] [PubMed]
35. Ding, Y.; Huang, L.N.; Barakat, T.; Su, B.L. A Novel 3DOM TiO_2 based multifunctional photocatalytic and catalytic platform for energy regeneration and pollutants degradation. *Adv. Mater. Interfaces* **2020**, *8*, 2001879. [CrossRef]
36. Cai, J.M.; Wu, M.Q.; Wang, Y.T.; Zhang, H.; Meng, M.; Tian, Y.; Li, X.G.; Zhang, J.; Zheng, L.R.; Gong, J.L. Synergetic enhancement of light harvesting and charge separation over surface-disorder engineered TiO_2 photonic crystals. *Chem* **2017**, *2*, 877–892. [CrossRef]
37. Zhao, H.; Li, C.F.; Hu, Z.Y.; Liu, J.; Li, Y.; Hu, J.G.; Tendeloo, G.V.; Chen, L.H.; Su, B.L. Size effect of bifunctional gold in hierarchical titanium oxide-gold-cadmium sulfide with slow photon effect for unprecedented visible-light hydrogen production. *J. Colloid Interface Sci.* **2021**, *604*, 131–140. [CrossRef]
38. Lin, J.; Liu, K.; Chen, X.F. Synthesis of periodically structured titania nanotube films and their potential for photonic applications. *Small* **2011**, *7*, 1784–1789. [CrossRef]
39. Zhou, W.M.; Wang, J.; Wang, X.G.; Li, J.F.; Li, Y.; Wang, C.W. The preparation of high quality heterostructure photonic crystals by using pulse anodization and their enhanced photocatalytic property. *Phys. E* **2019**, *114*, 113571. [CrossRef]
40. Yip, C.T.; Huang, H.T.; Zhou, L.M.; Xie, K.Y.; Wang, Y.; Feng, T.H.; Li, J.S.; Tam, W.Y. Direct and seamless coupling of TiO_2 nanotube photonic crystal to dye-sensitized solar cell: A single-step approach. *Adv. Mater.* **2011**, *23*, 5624–5628. [CrossRef]
41. Guo, M.; Xie, K.Y.; Lin, J.; Yong, Z.H.; Yip, C.T.; Zhou, L.M.; Wang, Y.; Huang, H.T. Design and coupling of multifunctional TiO_2 nanotube photonic crystal to nanocrystalline titania layer as semi-transparent photoanode for dye-sensitized solar cell. *Energy Environ. Sci.* **2012**, *5*, 9881. [CrossRef]
42. Meng, M.; Yang, L.; Yang, J.; Zhu, Y.; Li, C.Y.; Xia, H.J.; Yuan, H.L.; Zhang, M.; Zhao, Y.; Tian, F.S.; et al. Two-dimensional lateral anatase-rutile TiO_2 phase junctions with oxygen vacancies for robust photoelectrochemical water splitting. *J. Colloid Interface Sci.* **2023**, *648*, 56–65. [CrossRef] [PubMed]
43. Sun, X.H.; Zhao, X.Y.; Zhang, X.Y.; Wu, G.H.; Rong, X.H.; Wang, X.W. TiO_2 Nanosheets/$Ti_3C_2T_x$ MXene 2D/2D composites for excellent microwave absorption. *ACS Appl. Nano Mater.* **2023**, *6*, 14421–14430. [CrossRef]
44. Li, J.F.; Wang, J.; Wang, X.T.; Wang, X.G.; Li, Y.; Wang, C.W. Bandgap engineering of TiO_2 nanotube photonic crystals for enhancement of photocatalytic capability. *CrystEngComm* **2020**, *22*, 1929–1938. [CrossRef]
45. Meng, M.; Li, C.Y.; Li, J.T.; Wu, J.; Feng, Y.M.; Sun, L.L.; Yuan, H.L.; Liu, K.L. 3D TiO_2 nanotube arrays with anatase-rutile phase junction and oxygen vacancies for promoted photoelectrochemical water splitting. *J. Phys. D Appl. Phys.* **2023**, *56*, 055502. [CrossRef]
46. Low, J.X.; Zhang, L.Y.; Zhu, B.C.; Liu, Z.Y.; Yu, J.G. TiO_2 photonic crystals with localized surface photothermal effect and enhanced photocatalytic CO_2 reduction activity. *ACS Sustain. Chem. Eng.* **2018**, *6*, 15653–15661. [CrossRef]

47. Meng, M.; Feng, Y.M.; Li, C.Y.; Gan, Z.X.; Yuan, H.L.; Zhang, H.H. Black 3D-TiO$_2$ nanotube arrays on Ti meshes for boosted photoelectrochemical water splitting. *Nanomaterials* **2022**, *12*, 1447. [CrossRef]
48. Guo, M.; Yong, Z.H.; Xie, K.Y.; Lin, J.; Wang, Y.; Huang, H.T. Enhanced light harvesting in dye-sensitized solar cells coupled with titania nanotube photonic crystals: A theoretical study. *ACS Appl. Mater. Interfaces* **2013**, *5*, 13022–13028. [CrossRef]
49. Shi, L.; Zhou, W.; Li, Z.; Koul, S.; Kushima, A.; Yang, Y. Periodically ordered nanoporous perovskite photoelectrode for efficient photoelectrochemical water splitting. *ACS Nano* **2018**, *12*, 6335–6342.200A. [CrossRef]
50. Li, P.; Chen, S.L.; Wang, A.J.; Wang, Y. Probing photon localization effect between titania and photonic crystals on enhanced photocatalytic activity of titania film. *Chem. Eng. J.* **2016**, *284*, 305–314. [CrossRef]
51. Chen, S.L.; Wang, A.J.; Dai, C.; Benziger, J.B.; Liu, X.C. The effect of photonic band gap on the photo-catalytic activity of ncTiO$_2$/SnO$_2$ photonic crystal composite membranes. *Chem. Eng. J.* **2014**, *249*, 48–53. [CrossRef]
52. Li, P.; Wang, Y.; Chen, S.L.; Wang, A.J. Enhancement of gas-solid photocatalytic activity of nanocrystalline TiO$_2$ by SiO$_2$ opal photonic crystal. *J. Mater. Sci.* **2016**, *51*, 2079–2089. [CrossRef]

Disclaimer/Publisher's Note: The statements, opinions and data contained in all publications are solely those of the individual author(s) and contributor(s) and not of MDPI and/or the editor(s). MDPI and/or the editor(s) disclaim responsibility for any injury to people or property resulting from any ideas, methods, instructions or products referred to in the content.

Article

Tailored Triggering of High-Quality Multi-Dimensional Coupled Topological States in Valley Photonic Crystals

Guangxu Su [1,†], Jiangle He [1,†], Xiaofei Ye [2], Hengming Yao [1], Yaxuan Li [1], Junzheng Hu [2], Minghui Lu [3], Peng Zhan [2,*] and Fanxin Liu [1,*]

1. Department of Applied Physics, Zhejiang University of Technology, Hangzhou 310023, China; gxsu@zjut.edu.cn (G.S.); 2112109030@zjut.edu.cn (J.H.); 211123090024@zjut.edu.cn (H.Y.); 211122090031@zjut.edu.cn (Y.L.)
2. National Laboratory of Solid State Microstructures, Collaborative Innovation Center of Advanced Microstructures, School of Physics, Nanjing University, Nanjing 210093, China; mf21220006@smail.nju.edu.cn (X.Y.); dg21220019@smail.nju.edu.cn (J.H.)
3. National Laboratory of Solid State Microstructures, Department of Materials Science and Engineering, Nanjing University, Nanjing 210093, China; luminghui@nju.edu.cn
* Correspondence: zhanpeng@nju.edu.cn (P.Z.); liufanxin@zjut.edu.cn (F.L.)
† These authors contributed equally to this work.

Citation: Su, G.; He, J.; Ye, X.; Yao, H.; Li, Y.; Hu, J.; Lu, M.; Zhan, P.; Liu, F. Tailored Triggering of High-Quality Multi-Dimensional Coupled Topological States in Valley Photonic Crystals. Nanomaterials 2024, 14, 885. https://doi.org/10.3390/nano14100885

Academic Editors: Detlef W. Bahnemann and Ze Don Kvon

Received: 14 March 2024
Revised: 13 May 2024
Accepted: 16 May 2024
Published: 19 May 2024

Copyright: © 2024 by the authors. Licensee MDPI, Basel, Switzerland. This article is an open access article distributed under the terms and conditions of the Creative Commons Attribution (CC BY) license (https://creativecommons.org/licenses/by/4.0/).

Abstract: The combination of higher-order topological insulators and valley photonic crystals has recently aroused extensive attentions due to the great potential in flexible and efficient optical field manipulations. Here, we computationally propose a photonic device for the 1550 nm communication band, in which the topologically protected electromagnetic modes with high quality can be selectively triggered and modulated on demand. Through introducing two valley photonic crystal units without any structural alteration, we successfully achieve multi-dimensional coupled topological states thanks to the diverse electromagnetic characteristics of two valley edge states. According to the simulations, the constructed topological photonic devices can realize Fano lines on the spectrum and show high-quality localized modes by tuning the coupling strength between the zero-dimensional valley corner states and the one-dimensional valley edge states. Furthermore, we extend the valley-locked properties of edge states to higher-order valley topological insulators, where the selected corner states can be directionally excited by chiral source. More interestingly, we find that the modulation of multi-dimensional coupled photonic topological states with pseudospin dependence become more efficient compared with those uncoupled modes. This work presents a valuable approach for multi-dimensional optical field manipulation, which may support potential applications in on-chip integrated nanophotonic devices.

Keywords: optical microcavity; multi-dimensional coupled topological states; higher-order photonic topological insulators; valley photonic crystals; pseudospin dependence

1. Introduction

The discovery of Chern insulators and a series of proposed topological effects in condensed matter physics has driven the development of topological photonics [1–3], which brings new avenues for transmitting and localizing light [4,5]. Photonic crystals are analogs of conventional crystals that replace the atomic lattice with a periodic medium, providing an excellent platform for topological physics due to the controllability band structure [6]. In practice, defects and impurities are inevitably introduced in the sample preparation, leading to energy loss and signal distortion. In the face of the above difficulties, the topologically protected photonic states are proposed and demonstrated to be of benefit for the dissipationless transport dynamics of light [7–10]. More recently, photonic higher-order topological insulators (HOTIs) with bulk-edge-corner correspondences have been extensively studied for their ability to control light in multi-dimensions, in which the topological index can be characteristic by the vectored Zak phase and Wannier center [11–16]. In addition, by introducing coupling effects between a series of topological states, the quality of a nanocavity can be further improved, which also brings extra freedom to manipulate light [17–20]. However, the photonic topological edge states and lower-dimensional corner states in HOTIs tend to be discrete in spectrum. To realize multi-dimensional coupling, the structure needs to introduce unit distortion or more complex artificial design, which would narrow the bandgap of bulk and limit the development of related applications.

Valley photonic crystals (VPCs) with non-zero Berry curvature in momentum space [21–23] provide a new method to realize higher-order topological phases, which has already been successfully demonstrated in many lattice structures such as kagome lattices [24], triangular lattices [25], honeycomb lattices [26], and square lattices [27]. Among them, the supported valley corner states (VCSs) are robust and valley-locked-dependent [28–30]. Based on this feature, several interesting optical devices have been designed, such as topological all-optical switches [31] and topological rainbows [32]. By combining HOTIs and valley freedom, the structure can both support diverse types of topological states, which may support potential applications in topological lasers, topological optical switches, and on-chip integrated optical circuits.

In this work, we computationally propose an on-chip photonic device for the 1550 nm communication band, in which the multi-dimensional coupled topological states are achieved with two types of VPCs unit. By optimizing the coupling strength between the zero-dimensional VCSs and the one-dimensional valley edge states (VESs), the transmission spectrum presents a typical Fano line, showing as a high-quality localized mode. Furthermore, we extend the valley Hall effect of light to a high-order version, and successfully visualize the directed excitations of coupled VCSs with different chiral sources. The simulated results provide a versatile way to manipulate light based on VCSs and VESs, which can also extend to other electromagnetic wave ranges by adjusting the structure size. By the way, the valuable approach for multi-dimensional optical field manipulation can extend to other material systems, such as GaAs or InP, as long as we replace the corresponding refractive index parameters and fine-tune the structure parameters.

2. Results and Discussion

The designed VPC sample is a silicon on insulator (SOI) with a 220 nm thick silicon layer and specific periodic holes. As shown in Figure 1a, the valley photonic structure is arranged in a honeycomb lattice with $a = 470$ nm period, and there are two rounded equilateral triangular air holes with side lengths of l_1 and l_2 in a single cell, where $\delta = l_1 - l_2$ and $l_1 + l_2 = a$. For triangular holes, VPC has a larger band gap compared to the circular holes. And the effective refractive index of silicon is defined as 2.83. In all numerical calculations, we use the commercial software COMSOL Multiphysics 5.4 based on the finite element method. The periodic boundary conditions and scattering boundary conditions are used for corresponding interfaces. And the mesh is set up as the build-in physical field segmentations. The calculated band structures of VPCs under transverse electric (TE) polarization are as shown in Figure 1. When $\delta = 0$, the VPC has C_{6v} symmetry and the corresponding band structure presents an obvious degenerate Dirac point at the K(K') valley. When $\delta \neq 0$, the VPC changes to C_3 symmetry and the Dirac cone can be gapped out, leading to two valley states with opposite circularly polarized chirality at the two unequal K(K') valleys. For $l_1 > l_2$, the K valley state in the first band is a right-handed circularly polarized (RCP) mode and the K valley state in the second band is a left-handed circularly polarized (LCP) mode. For $l_1 < l_2$, the two chiral polarizations of the K valley state are inverted, indicating the topological phase transition. Here, we designed two VPC units with $\delta = 0.6a$ (VPC_1) and $\delta = -0.6a$ (VPC_2), which have the same band structure but different topological phases. Furthermore, we numerically calculated the Berry curvature of the first band:

$$\Omega_n(\boldsymbol{k}) = \nabla_{\boldsymbol{k}} \times A_n(\boldsymbol{k}) = \frac{\partial A_y(\boldsymbol{k})}{\partial k_x} - \frac{\partial A_x(\boldsymbol{k})}{\partial k_y}$$

where $A_n(\mathbf{k}) = i\langle u_{n,\mathbf{k}}|\nabla_{\mathbf{k}}|u_{n,\mathbf{k}}\rangle$ is the Berry connection and $|u_{n,\mathbf{k}}\rangle$ is the Bloch periodic function. Although the VPCs do not break the time-reversal symmetry, the Chern number of the band is zero and the valley Chern number at the $K(K')$ valley is nonzero due to the breaking of the space-reversal symmetry. As shown in Figure 1d, the Berry curvature of VPC_1 near the $K(K')$ valley is greater (less) than zero, while the Berry curvature of VPC_2 near the $K(K')$ valley is less (greater) than zero. For the same band, the VPC_1 and VPC_2 satisfy $\Omega(\mathbf{k}) = -\Omega(-\mathbf{k})$. The integral of the Berry curvature is calculated as shown below:

$$C_{K/K'} = \frac{1}{2\pi}\int_{HBZ_{K/K'}} \Omega_n(\mathbf{k})d^2\mathbf{k}$$

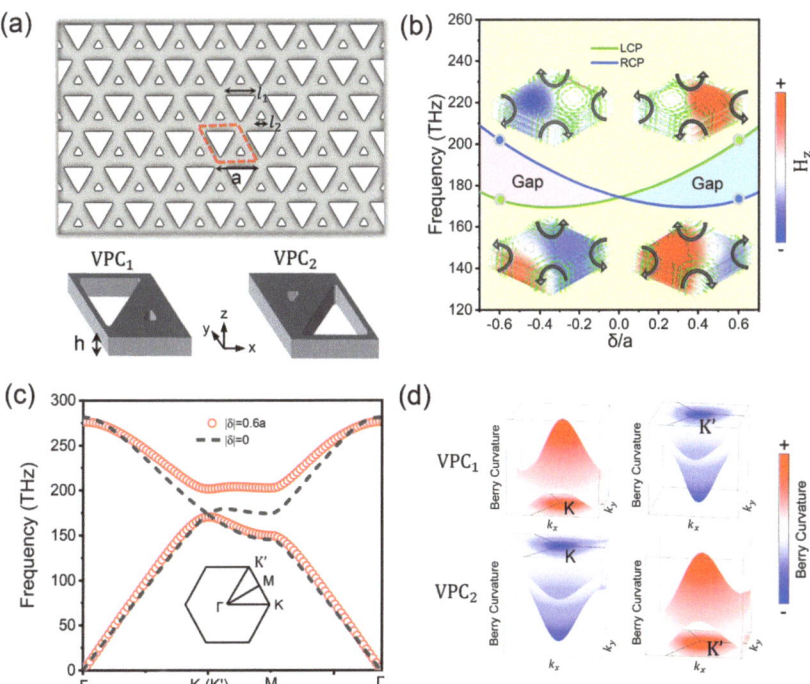

Figure 1. Topological phase transition and band structure. (**a**) Schematic of the VPC; the red dashed lines are the initial unit cell of the VPC. (**b**) A phase diagram showing the variation of the band gap as a function of δ. The inset shows the field distribution and Poynting vectors at selected points, shown with green and blue points; the rose and blue area corresponds to two opposite topological phases. (**c**) The band structures for VPCs. (**d**) Distribution of Berry curvature around K valley and K' valley for VPC_1 and VPC_2.

The valley Chern number of our system is half-integer, $C_K = -C_{K'} = 1/2$ for VPC_1 and $C_K = -C_{K'} = -1/2$ for VPC_2, indicating two opposite topological phases.

Next, we demonstrate two topological VESs based on the VPC_1 and VPC_2. For type I splicing interface, as shown in the left of Figure 2c, the larger triangular holes at the splicing interface are edge to edge. Meanwhile, for type II splicing interface, as shown in the right of Figure 2c, the larger triangular holes at the splicing interface are cusp to cusp. The projected band structures of two types of splicing structure are as shown in Figure 2; there is a bandgap between the type II VES (dashed line) and upper bulk states, which provides a possible coupling effect between VCS (orange dashed line) and another type I VES (solid line). In this case, the structure does not need to introduce any unit distortion or more complex artificial design. In addition, the valley-dependent VESs have opposite group velocities and vortices near the K valley and K' valley, which originates from the valley pseudospin-momentum locking. In order to visualize this physical feature, we calculated the two edge states at the K valley (labeled with different colored diamond symbols in Figure 2a) with the Poynting vectors and H_z phases as in Figure 2c. For the upper and lower interfaces of the splicing interface, the energy flow direction and phase vortex direction at K valley are opposite. And the dependencies for two types of VESs are also opposite, indicating the valley-momentum locking properties. Due to the time-reversal symmetry, the K' valley of the same structure has similar dependencies, as shown in Figure S1.

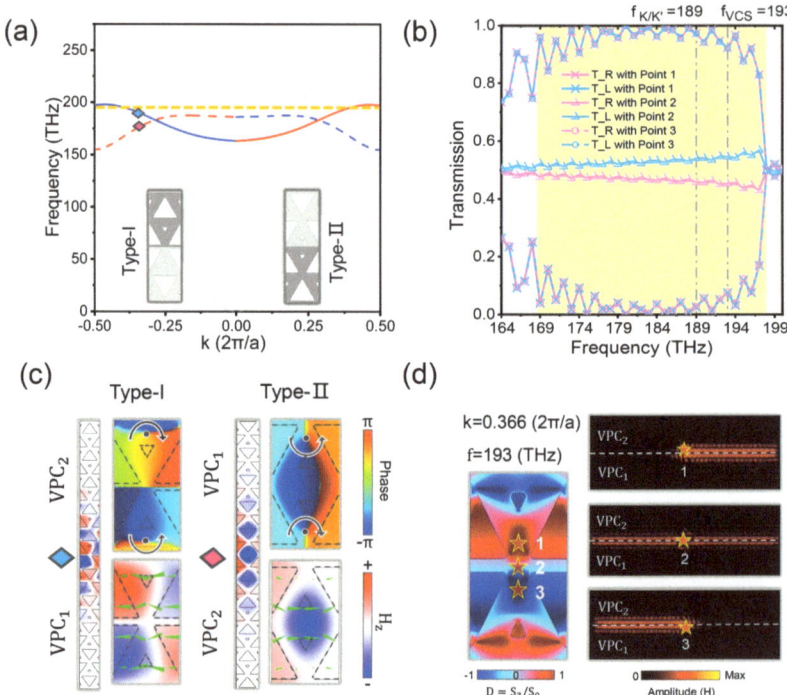

Figure 2. Topological projection band diagrams of VESs and valley-momentum locking phenomenon. (a) Band structure for the type I and type II interfaces; the solid and dashed line are the eigenmode of type I and II VES, respectively. And the grey area is bulk modes. (b) Normalized energy flow at the left or right ports when an RCP source is at positions 1, 2, and 3. (c) Valley-momentum locking properties of two types of VESs at K valley, which is selected from (a). (d) Positional dependence of the normalized Stokes S_3/S_0 parameter and the field distributions with the RCP source at positions 1, 2, and 3.

Based on the feature, we can achieve a directional transmission of light by using different circular polarizations of light. Here, we define the Stocks parameter to quantitatively analyze the unidirectional transmission ability of VESs as shown below [33,34]:

$$D = \frac{P_R - P_L}{P_R + P_L} = \frac{S_3}{S_0} = \frac{2Im(E_x^* E_y)}{|E_x|^2 + |E_y|^2}$$

Taking the type I VES as an example, $S_0 = |E_x|^2 + |E_y|^2$ and $S_3 = 2Im(E_x^* E_y)$ indicates the circular polarization point of the local polarization as the RCP (LCP), respectively. P_R

(P_L) represent the energy of light transmitting to the right (left) and $D = \pm 1$ represents that, ideally, light is transmitted completely to the right or left.

As shown in Figure 2b,d, we place an RCP source in three typical areas near the splicing interface; position 1 and 3 are the upper zone with $D = 1$ and the bottom zone with $D = -1$, respectively. Position 2 is the center of the interface with $D = 0$, which means that the unidirectional transmission of energy becomes worse. In simulation, we detected the energy of light at left or right ports to verify the directional transmission capability and the simulated field distributions agree well with the theoretical predictions. When an RCP source is placed at position 1, the energy of electromagnetic waves from frequency 164 to 198 THz will transmit to the right, where the light leaves through the right output port. And the energy will transmit to the left when the RCP source is placed at position 3 or is replaced by LCP source. When the RCP source is put to position 2, the transmission of light to both sides is almost equal.

The valley-momentum locking mentioned above can be understood from the quantized valley Chern number of the VESs. For the type I VES, the valley Chern number can be defined as $C_{IK} = C_{VPC_2K} - C_{VPC_1K} = -0.5 - 0.5 = -1, C_{IK'} = C_{VPC_2K'} - C_{VPC_1K'} = 1$. Similarly, the valley Chern number of the type II edge state can be defined as $C_{IIK} = C_{VPC_1K} - C_{VPC_2K} = 0.5 - (-0.5) = 1$, $C_{IIK'} = C_{VPC_1K'} - C_{VPC_2K'} = -1$. Valley Chren numbers with the same sign have consistent valley dependence properties and vice versa. More importantly, the quantizable valley-momentum locking properties of VESs and the naturalness of coupling with VCSs are the keys to the next realization of nanocavities with high responsiveness and high performance.

By combining the HOTI and valley degrees of freedom, the splicing corners of two VPCs with different valley Chern numbers can excite the VCS modes due to the valley–valley interactions of the VESs. Here, we construct a trapezoidal splicing structure, where the VPC_1 unit is surrounded by VPC_2 unit. As shown in Figure 3a–c, although there are four splicing corners here, only two VCS can be support in the eigenmodes simulations. And the 60-degree splicing corner appears obvious, while 120-degree angle disappears. The selective activation of VCS is related to the sign flip of the valley Chern number at the splicing corners and more details can be found in our previous work [31]. Due to the collective coupling effects of two VCSs, the equivalent corners of the structure appear to be two asymmetric VCSs, leading to spectrum division. The bonding coupled VCS with lower frequency presents two synchronous nanocavities, named as $\varphi_C^+ = \varphi_1 + \varphi_1'$. Meanwhile, for the anti-bonding coupled VCS with higher frequency, the adjacent nanocavities present a π phase difference, named as $\varphi_C^- = \varphi_1 - \varphi_1'$. The physics behind this can be referred to the electrodynamics theory; when two electric dipoles in the same direction end to end are close to each other, the two dipoles attract each other to form a bond and the energy of the coupled system decreases, corresponding to the bonding coupled modes with lower eigenfrequency. Similarly, when two dipoles in the opposite direction end to end are close to each other, the two dipoles are mutually exclusive to form an anti-bond. Furthermore, we have extended the valley Hall effect of light to a high-order version, as shown in Figure 3d–f. When an RCP source is placed on the center of the splicing interface, the intensity at point B is slightly stronger than point A. The directional excitation of the nanocavity will be reversed once the LCP source is placed on the same position. This interesting phenomenon is related to the valley-momentum locking and pseudospin polarization of the two coupled VCSs. It is worth noting that, although there are two eigenvalues of the VCS in our system, only one resonant peak is observed, which might be attributed to the small difference between the two eigenfrequencies. Next, we have calculated the quality factor Q of the nanocavity, which can be expressed as $Q = \frac{\omega_0 \tau}{2} = \frac{\omega_0}{\Delta \omega}$ in terms of the resonance frequency (ω_0) and the decay time of the electromagnetic energy in the cavity (τ) or the resonance linewidth ($\Delta \omega$) and the normalized field-strength spectrum, which shows that the quality factor of the nanocavities are around 375. On the other hand, the energy conversion from source to the VCSs is also weak.

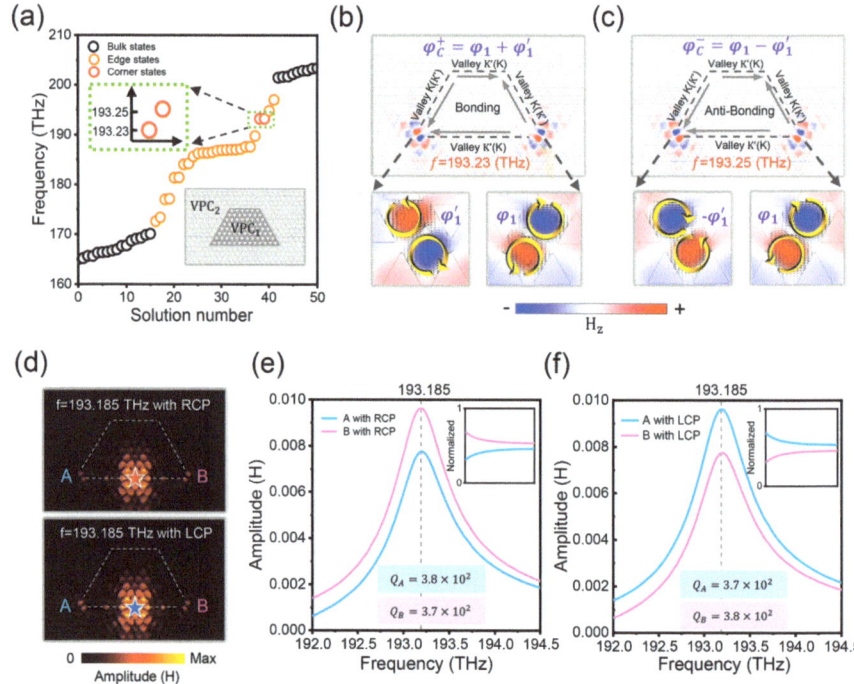

Figure 3. Pseudospin dependence of VCSs. (**a**) Eigenvalues of the bulk, edge, and corner states. (**b**,**c**) Field distribution of the coupled VCSs; arrows represent the direction of current. (**d**) Field distributions with different chiral sources at resonance frequencies, red (blue) stars are right-handed (left-handed) sources, respectively. (**e**,**f**) Field intensity at points A and B with an RCP or LCP source at the center position.

To further enhance the quality factor and response of VCSs, we have designed a waveguide-nanocavity coupled structure based on the VPC_1 and VPC_2, as shown in the inset of Figure 4a, where the zero-dimensional corner state and one-dimensional edge state can be naturally coupled. Here, the VES with an odd or even symmetric field distribution are defined as φ_E^+ or φ_E^-. Since the two wave functions with different symmetries are orthogonal to each other, $\left\langle \varphi_C^{+(-)} \middle| \varphi_E^{-(+)} \right\rangle = 0$, there are only four multi-dimensional coupled modes in our system, as shown in Figure 4. For $Mode_1 = \varphi_C^+ + \varphi_E^+$, the eigenfrequency is obviously lowest, corresponding to the bonding coupling between VES and VCS. And the eigenfrequency of $Mode_4 = \varphi_C^- - \varphi_E^-$ is highest, corresponding to the anti-bonding coupling between VES and VCS. As for the multi-dimensional coupled topological states with frequencies in between, the modes are defined as $Mode_2 = \varphi_C^- + \varphi_E^-$ and $Mode_3 = -\varphi_C^+ + \varphi_E^+$. These multi-dimensional coupled topological states can be distinguished from the field distributions, where the energy of $Mode_1$ and $Mode_4$ are mainly concentrated in the VESs, shown as bright modes, while the energy of $Mode_2$ and $Mode_3$ are mainly concentrated in the VCSs, shown as dark modes. By introducing coupling effects between the VCS and VES, the quality of nanocavities can be further improved, which also brings extra freedom to manipulate light on chip.

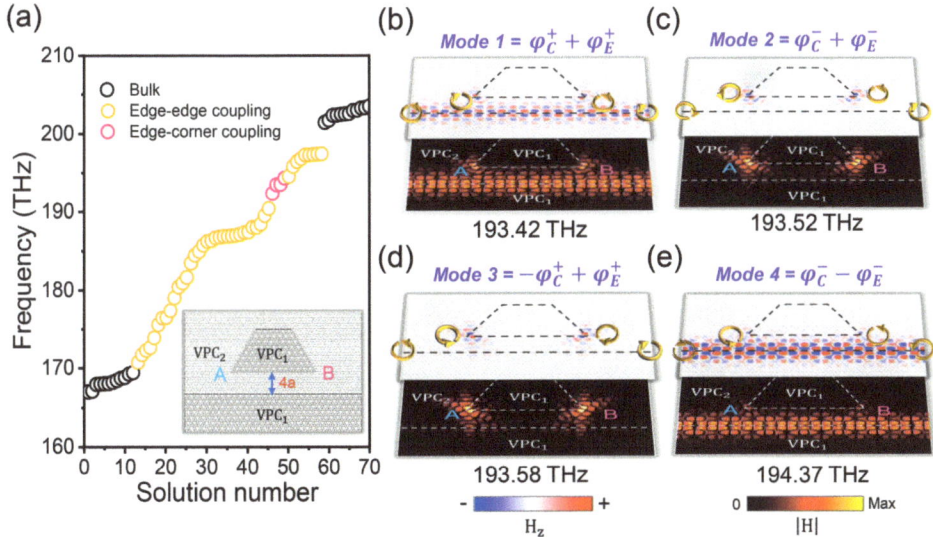

Figure 4. Characteristic of multi-dimensional coupled topological states. (**a**) Eigenvalues of the bulk, edge, and corner states; the inset shows a schematic structure. (**b**–**e**) Field distributions of four coupled topological states.

As shown in Figure 5, the quality factor and directional transmission capability of coupled VCSs have been significantly improved. When an RCP source is placed at position 1, the field intensity at the right corner is much stronger than the left corner in a wide wave range. If an LCP source is used, only the left and right corners are switched due to the time-reversal symmetry. When we only change the spatial position of the RCP source on the splice interfaces, the other structures are unchanged. And the results show that the field intensities of the right and left corner states are almost the same no matter which chiral source is placed at position 2. When RCP source is placed at position 3, the intensity of the left corner is much stronger than the right corners in a wide wave range. In the meantime, the intensity at corresponding corners is increased by 2–3 orders of magnitude compared to uncoupled VCS, as visualized in Figures 3 and 5. The reason for this high responsiveness and higher-order topological valley-locked characteristic is because the energy of the VCS at this point is directly affected by the selective coupling of the topological waveguide, and the responsiveness of the chiral source to the selective excitation of the VCS is equivalent to perturbation, i.e., the unidirectional transmission capability of the VES determines the field strength ratio at the two splice corners on both sides.

It is worth noting that, when the frequency of the excitation source is near the resonance frequency of the nanocavity, the field strengths on both sides of the splice corners are almost the same, which is due to the fact that the introduction of the nanocavity inevitably disrupts the overall symmetry of the lattice, and the unidirectional transmission ability of the VES is limited at the resonance frequency of the VCS, as shown in Figure 5e,g, where the topological waveguide still maintains a good unidirectional transmission capability, while the magnitude of the energy flow density on both sides is almost the same near the resonance frequency. In order to further characterize the excellent performance of the valley nanocavities, we calculate their quality factors, and the results show that both sides of the nanocavities have very high-quality factors in all three cases up to about 20,000, which is nearly 50 times higher than that of the previous uncoupled system.

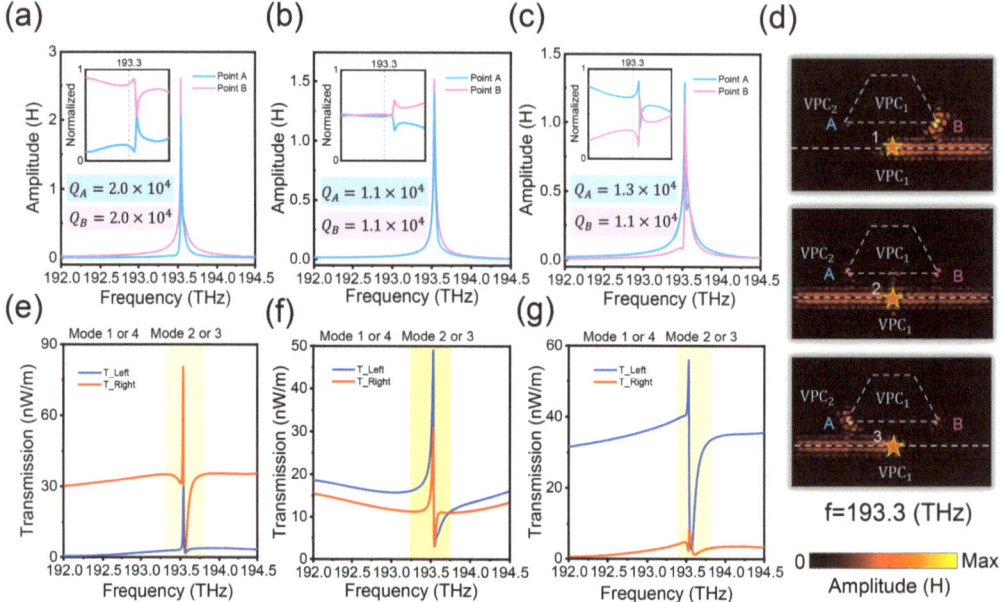

Figure 5. Pseudospin dependence of high-quality multi-dimensional coupled topological states. (**a**–**c**) Field intensities of the VCS nanocavities at points A and B when an RCP source is at positions 1, 2, and 3, respectively, and the inset shows the normalized field intensities. (**d**) Field distributions of the RCP source at the 1, 2, and 3 positions. (**e**–**g**) Transmission spectrum at the left or right ports when an RCP source is at positions 1, 2, and 3.

For the well-designed topologically protected nanophotonic devices with high responsiveness and high performance, the system also shows a tunable asymmetric spectral line in the transmission spectrum. And the physical mechanism behind it is the Fano resonance phenomenon arising from coherent interference between the discrete coupled VCS and the continuous VES near the resonance frequency, which requires that the resonance frequency of the discrete state is in the frequency range of the continuous state. Unlike other schemes, our structure naturally satisfies this condition without changing any parameters. As shown in Figure 5e–g, when we place the RCP source in positions 1 or 3, the transmission spectra at the left port present a typical Fano line shape, while the transmission spectra at the right port shows an electromagnetically induced transparency-like (EIT-like) line shape. If the source is placed at position 2, the transmission spectra of the ports on both sides are EIT-like line shapes. By considering the field distribution at different wavelengths, we can intuitively understand the above spectral response as shown in Figure S2. This reveals that the Fano and EIT-like resonance phenomenon originates from the coherent interference between the edge–corner coupling states with different line widths, where the modes on the wide transmission spectral lines are all $Mode_1$ or $Mode_4$ with energy concentrated in the topological waveguide, corresponding to the bright modes, and the modes on the narrow spectral lines are all $Mode_2$ or $Mode_3$ with energy concentrated in the valley nanocavity, corresponding to the dark modes. The Fano resonance is formed by the destructive and constructive interference of the two modes. On the other hand, since the system maintains the time-reversal symmetry, when we place the LCP source at different positions of the splicing line, the result is only that the spectral lines are swapped, while the other laws remain the same, as shown in Figure S3. In addition to the advantage of tunability, our system is also extremely robust. As shown in Figure S4, where we destroy the geometry near the nanocavity by replacing it from the original small nanopore to a large one, the calculation results show that only the resonance frequency of the cavity is slightly blue shifted, while the cavity's localization, quality factor, and higher-order valley-locking properties are basically unaffected, and the Fano resonance spectral lines are also relatively well protected. In conclusion, this system and modulation we have established not only greatly improves the quality factor of the valley nanocavity and the responsiveness of the higher-order valley-locking properties but also achieves topologically protected tunable Fano and EIT-like resonance spectra in the same structure.

Finally, we also demonstrate the modulation of the system by changing the coupling distance between the cavity and the waveguide. Figure 6a,c shows the evolution of the quality factors of the left and right cavities with the coupling distance when the right-handed chiral source is in the splicing line at positions 1 and 2, and Figure 6b,d show the transmission spectra corresponding to the left and right ends. The relevant data when the RCP source is at position 3 are shown in Figure S5. These results show that, as the coupling strength decreases, the quality factor of the cavities increases roughly linearly, up to about 60,000. From the transmission spectra, we can derive that the resonance frequencies of the dark modes are gradually blue-shifted, and the peaks of the Fano spectral line and the EIT-like spectral line are also reduced substantially; however, their shapes are basically maintained in the same way. In addition to the coupling distance of four cell sizes, the transmission spectra of the RCP source are all classical Fano lines when placed at positions 1 and 3 and the transmission spectra are all EIT-like lines when placed at position 2, which provides a new method and path for the modulation of the topological Fano transmission spectral lines.

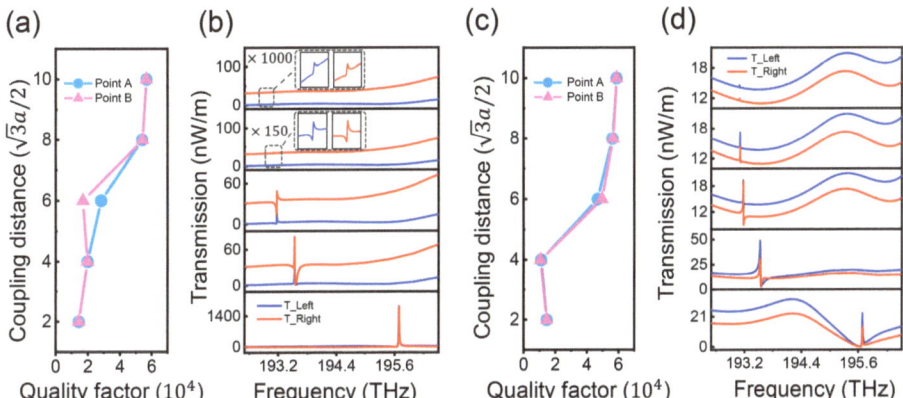

Figure 6. Regulation of multi-dimensional coupled topological states with different coupling strength. (**a**,**b**) Dependence of the quality factor and transmission spectrum of the structure with different coupling strength when the RCP source is at position 1. (**c**,**d**) Corresponding parameters when the RCP source is at position 2.

3. Conclusions

In summary, we computationally propose a topologically protected high-quality optical nanocavity, which can be selectively triggered and modulated on demand. Based on the mismatch in the spectrum of two valley edge states, we successfully demonstrate the coupling effect between the zero-dimensional valley corner states and the one-dimensional valley edge states without any structural alteration. By optimizing the coupling strength between the valley corner states and edge states, we observe an extremely high-quality localized mode. Furthermore, we have extended the valley Hall effect of light to a higher-order version, where the selected photonic topological corner states can be directionally excited with different polarizations of light and the coupled VCS with pseudospin dependence become more efficient. This work visualizes an efficient and flexible electromagnetic mode with pseudospin dependence, which is valuable for the development of on-chip integrated topological photonic devices.

Supplementary Materials: The following supporting information can be downloaded at: https://www.mdpi.com/article/10.3390/nano14100885/s1, Figure S1: Valley-momentum locking at the K' valley. Figure S2: Field distribution of different hybrid modes. Figure S3: Excitation of coupled topological states with LCP source. Figure S4: Robustness of coupled topological states. Figure S5: Effect of coupling strength on coupled topological states when RCP source is at position 3.

Author Contributions: Formal analysis, X.Y., H.Y., Y.L. and J.H. (Junzheng Hu); Data curation, G.S. and J.H. (Jiangle He); Writing—original draft, J.H. (Jiangle He); Writing—review & editing, G.S.; Project administration, M.L., P.Z. and F.L. All authors have read and agreed to the published version of the manuscript.

Funding: This work is funded by the National Key Research and Development Program of China (Nos. 2022YFA1404302), National Natural Science Foundation of China (Nos. 11974015, 12304431, 12174189), and Natural Science Foundation of Zhejiang Province (Nos. LQ24A040015, LZ22A040008).

Data Availability Statement: The data that support the findings of this study are available from the corresponding author upon reasonable request.

Conflicts of Interest: The authors declare no conflicts of interest.

References

1. Hasan, M.Z.; Kane, C.L. Colloquium: Topological insulators. *Rev. Mod. Phys.* **2010**, *82*, 3045. [CrossRef]
2. Qi, X.L.; Zhang, S.C. Topological insulators and superconductors. *Rev. Mod. Phys.* **2011**, *83*, 1057. [CrossRef]
3. Fu, L. Topological crystalline insulators. *Phys. Rev. Lett.* **2011**, *106*, 106802. [CrossRef] [PubMed]
4. Lu, L.; Joannopoulos, J.D.; Soljačić, M. Topological photonics. *Nat. Photonics* **2014**, *8*, 821–829. [CrossRef]
5. Ozawa, T.; Price, H.M.; Amo, A.; Goldman, N.; Hafezi, M.; Lu, L.; Rechtsman, M.C.; Schuster, D.; Simon, J.; Zilberberg, O.; et al. Topological photonics. *Rev. Mod. Phys.* **2019**, *91*, 015006. [CrossRef]
6. Yablonovitch, E. Inhibited spontaneous emission in solid state physics and electronics. *Phys. Rev. Lett.* **1987**, *58*, 2059. [CrossRef] [PubMed]
7. Hafezi, M.; Mittal, S.; Fan, J.; Migdall, A.; Taylor, J.M. Imaging topological edge states in silicon photonics. *Nat. Photonics* **2013**, *7*, 1001. [CrossRef]
8. Khanikaev, A.B.; Shvets, G. Two-dimensional topological photonics. *Nat. Photon.* **2017**, *11*, 763. [CrossRef]
9. Khanikaev, A.B.; Fleury, R.; Mousavi, S.H.; Alu, A. Topologically robust sound propagation in an angular-momentum-biased graphene-like resonator lattice. *Nat. Commun.* **2015**, *6*, 8260. [CrossRef]
10. Weiner, M.; Ni, X.; Li, M.; Alù, A.; Khanikaev, A.B. Demonstration of a third-order hierarchy of topological states in a three-dimensional acoustic metamaterial. *Sci. Advances* **2020**, *6*, 4166. [CrossRef]
11. Xie, B.Y.; Su, G.X.; Wang, H.F.; Su, H.; Shen, X.P.; Zhan, P.; Lu, M.H.; Wang, Z.L.; Chen, Y.F. Visualization of Higher-Order Topological Insulating Phases in Two-Dimensional Dielectric Photonic Crystals. *Phys. Rev. Lett.* **2019**, *122*, 233903. [CrossRef] [PubMed]
12. El Hassan, A.; Kunst, F.K.; Moritz, A.; Andler, G.; Bergholtz, E.J.; Bourennane, M. Corner States of Light in Photonic Waveguides. *Nat. Photonics* **2019**, *13*, 697. [CrossRef]
13. Li, M.; Zhirihin, D.; Gorlach, M.; Ni, X.; Filonov, D.; Slobozhanyuk, A.; Alù, A.; Khanikaev, A.B. Higher-order Topological States in Photonic Kagome Crystals with Longrange Interactions. *Nat. Photonics* **2020**, *14*, 89. [CrossRef]
14. Xie, B.Y.; Su, G.X.; Wang, H.F.; Liu, F.; Hu, L.; Yu, S.Y.; Zhan, P.; Lu, M.H.; Wang, Z.L.; Chen, Y.F. Higher-order quantum spin Hall effect in a photonic crystal. *Nat. Commun.* **2020**, *11*, 3768. [CrossRef] [PubMed]
15. Benalcazar, W.A.; Li, T.; Hughes, T.L. Quantization of fractional corner charge in Cn-symmetric higher-order topological crystalline insulators. *Phys. Rev. B* **2019**, *99*, 245151. [CrossRef]
16. Ezawa, M. Higher-order topological insulators and semimetals on the breathing Kagome and pyrochlore lattices. *Phys. Rev. Lett.* **2018**, *120*, 026801. [CrossRef] [PubMed]
17. Shi, A.Q.; Yan, B.; Ge, R.; Xie, J.L.; Peng, Y.C.; Li, H.; Sha, W.E.I.; Liu, J.J. Coupled cavity-waveguide based on topological corner state and edge state. *Opt. Lett.* **2021**, *46*, 1089–1092. [CrossRef] [PubMed]
18. Kang-Hyok, O.; Kim, K.H. Ultrahigh-Q Fano resonance using topological corner modes in second-order pseudospin-Hall photonic systems. *Opt. Laser Technol.* **2022**, *147*, 107616.
19. Xu, J.; Zang, X.F.; Zhan, X.D.; Liu, K.; Zhu, Y.M. Manipulating electromagnetic waves in a cavity-waveguide system with nontrivial and trivial modes. *Opt. Lett.* **2022**, *47*, 5204–5207. [CrossRef]
20. Gao, Y.F.; He, Y.H.; Maimaiti, A.; Jin, M.C.; He, Y.; Qi, X.F. Manipulation of coupling between topological edge state and corner state in photonic crystals. *Opt. Laser Technol.* **2022**, *155*, 108387. [CrossRef]

21. Ma, T.; Shvets, G. All-Si valley-hall photonic topological insulator. *New J. Phys.* **2016**, *18*, 025012. [CrossRef]
22. Chen, X.D.; Zhao, F.L.; Chen, M.; Dong, J.W. Valley-contrasting physics in all-dielectric photonic crystals: Orbital angular momentum and topological propagation. *Phys. Rev. B* **2017**, *96*, 020202. [CrossRef]
23. Dong, J.W.; Chen, X.D.; Zhu, H.Y.; Wang, Y.; Zhang, X. Valley photonic crystals for control of spin and topology. *Nat. Mater.* **2017**, *16*, 298–302. [CrossRef]
24. Wang, H.X.; Liang, L.; Jiang, B.; Hu, J.H.; Jiang, J.H. Higher-order topological phases in tunable C_3 symmetric photonic crystals. *Photonics Res.* **2021**, *9*, 1854–1864. [CrossRef]
25. Gao, Z.X.; Liao, J.Z.; Shi, F.L.; Shen, K.; Ma, F.; Chen, M.; Chen, X.D.; Dong, J.W. Observation of Unidirectional Bulk Modes and Robust Edge Modes in Triangular Photonic Crystals. *Laser Photonics Rev.* **2023**, *17*, 2201026. [CrossRef]
26. Shao, S.; Liang, L.; Hu, J.H.; Poo, Y.; Wang, H.X. Topological edge and corner states in honeycomb-kagome photonic crystals. *Opt. Express* **2023**, *31*, 17695–17708. [CrossRef] [PubMed]
27. Kim, K.H.; Om, K.K. Multiband Photonic Topological Valley-Hall Edge Modes and Second-Order Corner States in Square Lattices. *Adv. Opt. Mater.* **2021**, *9*, 2001865. [CrossRef]
28. Zhang, X.J.; Liu, L.; Lu, M.H.; Chen, Y.F. Valley-selective topological corner states in sonic crystals. *Phys. Rev. Lett.* **2021**, *126*, 156401. [CrossRef] [PubMed]
29. Zhou, R.; Lin, H.; Wu, Y.J.; Li, Z.F.; Yu, Z.H.; Liu, Y.; Xu, D.H. Higher-order valley vortices enabled by synchronized rotation in a photonic crystal. *Photonics Res.* **2022**, *10*, 1244–1254. [CrossRef]
30. Phan, H.T.; Liu, F.; Wakabayashi, K. Valley-dependent corner states in honeycomb photonic crystals without inversion symmetry. *Opt. Express* **2021**, *29*, 18277–18290. [CrossRef]
31. He, J.L.; Jia, S.Y.; Li, Y.X.; Hu, J.Z.; Huang, R.W.; Su, G.X.; Lu, M.H.; Zhan, P.; Liu, F.X. Selective activation of topological valley corner states in C_3-symmetric photonic crystals. *Appl. Phys. Lett.* **2023**, *123*, 031104. [CrossRef]
32. Liang, L.; Zhou, X.; Hu, J.H.; Wang, H.X.; Jiang, J.H.; Jiang, B. Rainbow trapping based on higher-order topological corner modes. *Opt. Lett.* **2022**, *47*, 1454–1457. [CrossRef] [PubMed]
33. Ruan, W.S.; He, X.T.; Zhao, F.L.; Dong, J.W. Analysis of unidirectional coupling in topological valley photonic crystal waveguides. *J. Light. Technol.* **2020**, *39*, 889–895. [CrossRef]
34. Lan, Z.H.; You, J.W.; Ren, Q.; Sha, W.E.I.; Panoiu, N.C. Second-harmonic generation via double topological valley-Hall kink modes in all-dielectric photonic crystals. *Phys. Rev. A* **2021**, *103*, 041502. [CrossRef]

Disclaimer/Publisher's Note: The statements, opinions and data contained in all publications are solely those of the individual author(s) and contributor(s) and not of MDPI and/or the editor(s). MDPI and/or the editor(s) disclaim responsibility for any injury to people or property resulting from any ideas, methods, instructions or products referred to in the content.

Article

Dual-Channel Switchable Metasurface Filters for Compact Spectral Imaging with Deep Compressive Reconstruction

Chang Wang [1,†], Xinyu Liu [1,†], Yang Zhang [1], Yan Sun [1], Zeqing Yu [1] and Zhenrong Zheng [1,2,*]

1 College of Optical Science and Engineering, Zhejiang University, Hangzhou 310027, China; changwang_optics@zju.edu.cn (C.W.); 11930054@zju.edu.cn (X.L.)
2 Intelligent Optics & Photonics Research Center, Jiaxing Research Institute, Zhejiang University, Jiaxing 314000, China
* Correspondence: zzr@zju.edu.cn
† These authors contributed equally to this work.

Citation: Wang, C.; Liu, X.; Zhang, Y.; Sun, Y.; Yu, Z.; Zheng, Z. Dual-Channel Switchable Metasurface Filters for Compact Spectral Imaging with Deep Compressive Reconstruction. *Nanomaterials* 2023, *13*, 2854. https://doi.org/10.3390/nano13212854

Academic Editor: Julian Maria Gonzalez Estevez

Received: 25 September 2023
Revised: 20 October 2023
Accepted: 26 October 2023
Published: 27 October 2023

Copyright: © 2023 by the authors. Licensee MDPI, Basel, Switzerland. This article is an open access article distributed under the terms and conditions of the Creative Commons Attribution (CC BY) license (https:// creativecommons.org/licenses/by/ 4.0/).

Abstract: Spectral imaging technology, which aims to capture images across multiple spectral channels and create a spectral data cube, has been widely utilized in various fields. However, conventional spectral imaging systems face challenges, such as slow acquisition speed and large size. The rapid development of optical metasurfaces, capable of manipulating light fields versatilely and miniaturizing optical components into ultrathin planar devices, offers a promising solution for compact hyperspectral imaging (HSI). This study proposes a compact snapshot compressive spectral imaging (SCSI) system by leveraging the spectral modulations of metasurfaces with dual-channel switchable metasurface filters and employing a deep-learning-based reconstruction algorithm. To achieve compactness, the proposed system integrates dual-channel switchable metasurface filters using twisted nematic liquid crystals (TNLCs) and anisotropic titanium dioxide (TiO_2) nanostructures. These thin metasurface filters are closely attached to the image sensor, resulting in a compact system. The TNLCs possess a broadband linear polarization conversion ability, enabling the rapid switching of the incidence polarization state between x-polarization and y-polarization by applying different voltages. This polarization conversion facilitates the generation of two groups of transmittance spectra for wavelength-encoding, providing richer information for spectral data cube reconstruction compared to that of other snapshot compressive spectral imaging techniques. In addition, instead of employing classic iterative compressive sensing (CS) algorithms, an end-to-end residual neural network (ResNet) is utilized to reconstruct the spectral data cube. This neural network leverages the 2-frame snapshot measurements of orthogonal polarization channels. The proposed hyperspectral imaging technology demonstrates superior reconstruction quality and speed compared to those of the traditional compressive hyperspectral image recovery methods. As a result, it is expected that this technology will have substantial implications in various domains, including but not limited to object detection, face recognition, food safety, biomedical imaging, agriculture surveillance, and so on.

Keywords: optical metasurface; hyperspectral imaging; deep learning

1. Introduction

Spectral imaging technology aims to capture images with multiple spectral channels, forming a spectral data cube essential for the identification, analysis, and classification of objects predicated on their distinctive spectral attributes. Hyperspectral imaging (HSI) technology has found extensive applications in remote sensing [1], medical diagnostics [2], object detection and recognition [3], agriculture surveillance [4], and other fields. However, the conventional spectral imaging methods, such as spatial scanning and spectral scanning, suffer from limitations, such as large volume and slow acquisition speed. To address these challenges, researchers have explored snapshot spectral imaging (SSI) [5–7] and compressive spectral imaging (CSI) [8–11] technologies. On the one hand, SSI systems have significantly improved the spectral image acquisition process. However, the early SSI methods faced limitations in obtaining a large number of spectral channels and relied on bulky optical systems for light splitting. On the other hand, CSI methods leverage compressive sensing (CS) theory [12,13] and encoding devices to block or filter the input light field, enabling the reconstruction of hyperspectral images with much fewer measurements than the number that Nyquist sampling requires. Further, snapshot compressive imaging (SCI) [6], a technique that captures high-dimensional (HD, $\geq$3D) data using a two-dimensional (2D) detector in very few shots, has been employed in spectral imaging, known as snapshot compressive spectral imaging (SCSI). SCSI was initially introduced as a dual-disperser coded aperture snapshot spectral imaging (DD-CASSI) [14] system, integrating compressive sensing into hyperspectral imaging. Numerous adapted hyperspectral imaging systems based on CASSI have emerged, including single-disperser CASSI (SD-CASSI) [15], multi-frame CASSI [16], dual-camera CSI [17], color-coded aperture CSI [18], and spatial–spectral encoded CSI [19]. However, the bulkiness and complexity of CASSI systems result in non-linear dispersion that necessitates calibration, leading to the degradation of spatial information and suboptimal recovery outcomes. Additionally, the energy loss of light and intricate optical components within the large system volume make CASSI frameworks impractical for portable applications. To overcome these limitations, phase-coded spectral imaging [20] has been developed to improve light throughput and reduce system volume, while wavelength-coded methods [20] have been pursued to achieve accurate and fast RGB-to-spectra recovery. Wavelength-coded approaches utilize RGB or broadband optical filters that can be extended to multiple designed broadband filters for precise wavelength encoding. In terms of spectral data cube decoding, CSI reconstruction algorithms can be categorized into model-based methods and learning-based methods [21]. The traditional iterative reconstruction approaches utilize designed measurements of the encoding process and prior knowledge. Thereinto, CS optimization algorithms, such as the two-step iterative shrinkage/thresholding (TwIST) algorithm [22], and prior conditions, like total variation (TV) regularization [23], have been introduced. Additionally, methods like basis function fitting [24] and dictionary learning [25] have been developed. However, these classic iterative algorithms often require long computation times and prior knowledge, resulting in limited reconstruction quality and applicability in mobile systems with speed requirements. With the fast development of planar optical elements as well as deep-learning algorithms, the compactness, reconstruction speediness, and quality of SCSI systems could be further improved.

For one thing, metasurfaces have gained considerable attention in recent years due to their ability to manipulate the incident wavefront versatilely with subwavelength resolution, offering control over amplitude [26–29], phase [30], polarization [31–35], and spectrum [36,37]. In addition, they can effectively miniaturize optical elements into compact, planar, and ultrathin devices. Thus, metasurfaces are suitable to be applied in spectral imaging systems [38,39] by functioning as broadband filters with diversified transmission spectra and working as compact and cost-efficient wavelength-coded apertures [40,41]. Furthermore, tunable and multifunctional metasurfaces provide greater design freedoms [42–46] with liquid crystals (LCs) being widely used for implementing tunable metasurfaces due to their unique birefringence properties and mature control mechanisms [45–52]. These characteristics possess significant potential in the implementation of compact, efficient, and precise SCSI systems.

In addition, deep-learning algorithms have emerged as an alternative for learning spatial–spectral priors and spectral reconstruction. They offer faster and more accurate reconstruction compared to iterative approaches, thanks to the strong fitting ability of deep-learning models that alleviates the high computation costs. Consequently, various deep neural network architectures, such as autoencoders [53], convolutional neural networks [54–58], generative adversarial networks [59], transformers [60], and others, have been utilized for spectral reconstruction. Additionally, the entire reconstruction process can be substituted with a neural network, and end-to-end (E2E) reconstruction allows for the sending of measurements into a deep neural network that directly outputs the reconstruction results. Specifically, residual neural networks (ResNet) [61,62], composed of convolutional layers with skip connections, have achieved outstanding performance in computer vision tasks, such as image classification and object detection, making them suitable for spectral reconstruction purposes.

This study proposes a novel snapshot compressive hyperspectral imaging system called MD-SCSI, which is based on dual-channel switchable metasurface filters and a deep-learning-empowered compressive reconstruction algorithm. MD-SCSI incorporates two key innovations: an SCSI hardware encoder that utilizes dual-channel switchable metasurface filters and a deep compressive reconstruction algorithm that employs E2E convolutional neural networks based on the ResNet architecture. On the one hand, the dual-channel switchable metasurface filters are constructed by integrating twisted nematic liquid crystals (TNLCs) with all-dielectric metasurfaces composed of anisotropic titanium dioxide (TiO_2) meta-atoms. These thin metasurface filters are tightly integrated onto the image sensor, resulting in a compact system design. The TNLCs possess a broad linear polarization conversion capability [52], allowing for the rapid switching of the incidence polarization state between x-polarization and y-polarization by applying different voltages. This capability allows for the metasurface filters to operate as broadband optical filter arrays that are sensitive to polarization, thereby offering two distinct sets of transmittance spectra for wavelength encoding. Consequently, they facilitate the generation of 2-frame snapshots that capture spectral information, which in turn benefits the reconstruction of spectral data cubes. On the other hand, an E2E ResNet, instead of the traditional iterative CS algorithms, is employed to achieve hyperspectral imaging reconstruction. The ResNet is trained on a synthetic dataset using the 2-frame snapshot measurements obtained from the orthogonal polarization channels of MD-SCSI. By conducting a comparative analysis of the HSI reconstruction outcomes achieved using MD-SCSI against alternative methods, including a classical DD-CASSI method using random coded apertures and iterative CS algorithms with dictionary-learning based recovery (CASSI-DBR), method replacing the algorithm of MD-SCSI with iterative CS algorithms with dictionary-learning-based recovery (Meta-DBR), DD-CASSI using deep neural networks for recovery (CASSI-Net), and reconstruction with either the x-linear or y-linear polarization output channel of MD-SCSI (x-pol or y-pol). The precision and speediness of MD-SCSI have shown that MD-SCSI is superior to other methods, according to the improved reconstruction quality as well as speed. Generally, this research presents several contributions outlined as follows:

1. MD-SCSI realizes transmission spectra control by use of metasurfaces, employing two multiplexing input channels that work for orthogonally linear polarized light that can be rapidly switched using TNLCs. Additionally, the arrangement of meta-atoms within the metasurface units is optimized for minimization of coherence.
2. MD-SCSI achieves a compact SCSI framework rather than using additional elements or strategies of spatial-multiplexing, ensuring high-quality reconstruction while maintaining spatial resolution.
3. MD-SCSI enables fast and accurate HSI reconstruction by leveraging an end-to-end ResNet that is specially optimized for the dual-channel switchable metasurface filters, characterized by simplicity, high performance, rapid convergence, and exceptional generalization.

2. Materials and Methods

The proposed MD-SCSI system, as illustrated in Figure 1, consists of a vertically stacked image sensor, a layer of dual-channel switchable metasurface filters, and a layer of TNLCs. The metasurface layer is positioned between the sensor and the LC layer, and it comprises periodic micro-spectrometers. Each micro-spectrometer consists of 8×8 metasurface units with each unit measuring 3.45 μm $\times$ 3.45 μm, corresponding to a single complementary metal oxide semiconductor (CMOS) image sensor pixel. These metasurface units form a periodic array of 10×10 subwavelength anisotropic meta-atoms, which function as dual-channel switchable metasurface filters, exhibiting distinct transmission spectra for orthogonal polarization channels. The image sensor is integrated on top of the metasurface layer and accompanied by a micro lens array layer, while a thin TNLC cell is integrated beneath the metasurface layer to generate broadband linearly polarized incidences with orthogonal polarization states.

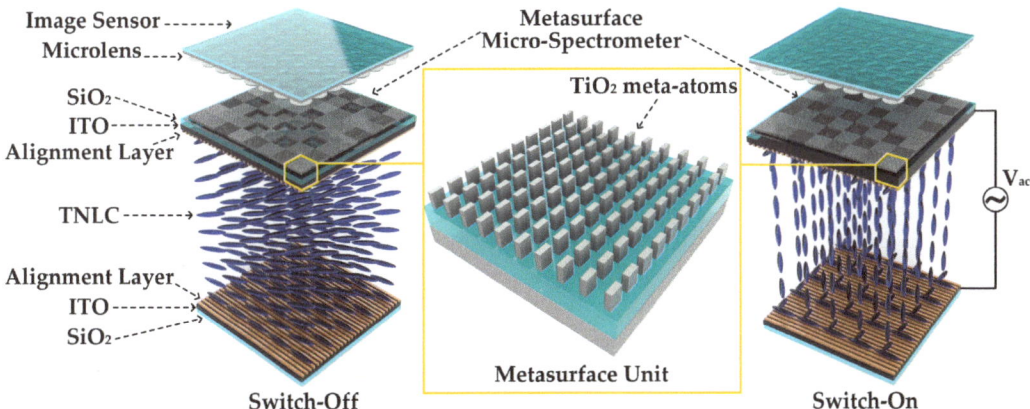

Figure 1. Schematic of the proposed MD-SCSI, consisting of a vertically stacked image sensor, a layer of dual-channel switchable metasurface filters with periodic micro-spectrometers, and a layer of TNLCs for tunable polarization conversion, enabling polarization channel selection of metasurface-filter arrays.

2.1. Metasurface Design

The device primarily consists of vertically stacked metasurface and TNLC layers. The upper metasurface layer is fabricated using high-ratio birefringent TiO_2 meta-atoms [63] with varying cross-sectional shapes on top of a quartz (silicon dioxide, SiO_2) substrate [64] (Figure 2). TiO_2 naturally has an exceptionally low extinction coefficient (k), a large refractive index (n), and high transmittance in the visible range, making the energy of the light strongly confined within each meta-atom, and its negligible extinction coefficient across the visible spectrum avoids Ohmic loss. In addition, the proposed metasurface filters achieve dual-channel wavelength-coding by the anisotropic meta-atoms, which performs very diversified transmission spectra for x-pol or y-pol incidences with great design flexibility of its shape, size, arrangements, and so on. Further, this metasurface-based architecture implements an ultrathin and compact system with the potential of cost-effective mass production [65]. The underlying layer comprises a thin TNLC cell filled with the commonly used LC material 4-cyano-4′-pentylbiphenyl (5CB) [49,66], which includes two orthogonally oriented alignment layers to pre-align the LC molecules. The LC cell was sandwiched between two 10 nm-thick indium tin oxide (ITO) [67] layers. The proposed metasurface, integrated with TNLCs in the visible region, is demonstrated in Figure 3. By applying different voltages to the TNLC cell, the orientation of the LC molecules can be adjusted, enabling the conversion of linearly polarized light. In the absence of an electric field, vertically incident linearly polarized light with a polarization direction parallel to the first alignment layer undergoes a 90° deflection along the twisting direction of the LC molecules while passing through the TNLC cell. However, when a voltage that exceeds the threshold voltage is applied to the TNLC, the long axes of the LC molecules begin to incline in the direction of the electric field [48,52]. With the exception of the molecules near the alignment layers, the long axes of the remaining molecules tend to rearrange parallel to the electric field [48], leading to the suppression of LC molecule polarization conversion. Importantly, this optical rotation of the LC molecules is wavelength-independent, enabling the TNLC cell to operate over a broadband range.

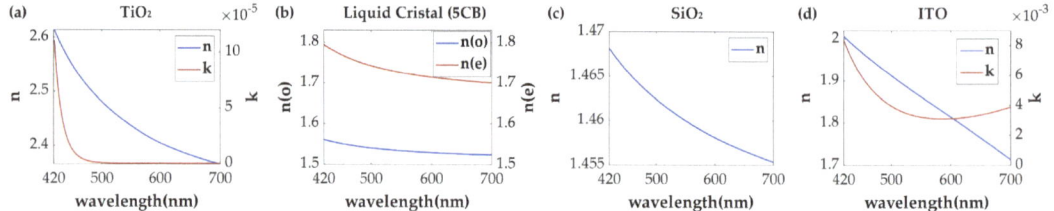

Figure 2. Refractive indices of (**a**) TiO$_2$ [63], (**b**) TNLC [66], (**c**) SiO$_2$ [64], and (**d**) ITO [67] materials used for constructing the meta-atoms.

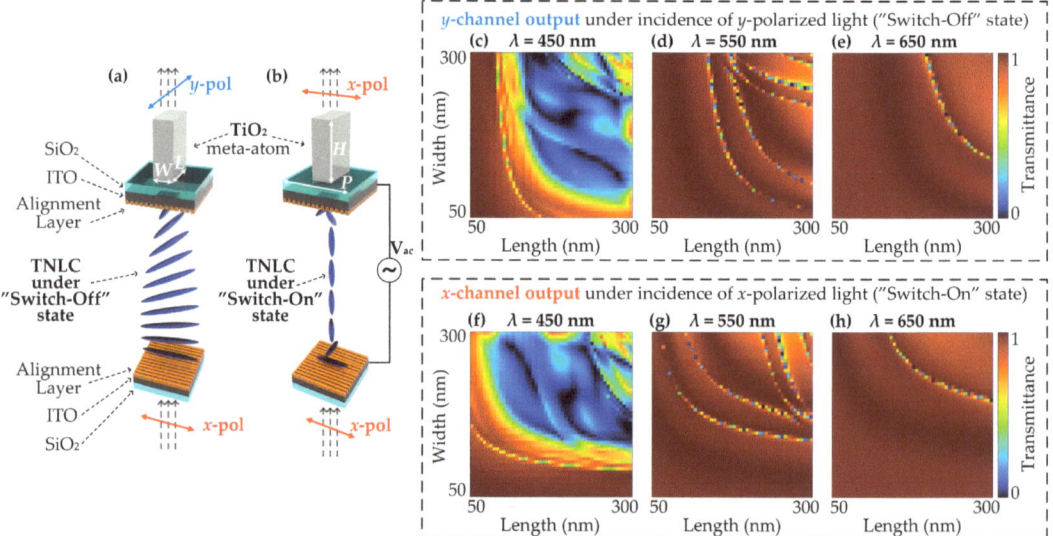

Figure 3. The detailed working principle of the proposed MD-SCSI. (**a**) The (helical) LC distribution in the 'Switch-Off' state under a meta-atom. (**b**) The situation in the 'Switch-On' state (no helical distribution) under a meta-atom. The polarization states of incident and output lights are indicated as red and blue arrows for x-polarization and y-polarization, respectively. (**c**–**e**) Diagram of transmittance response for different structural parameters of the meta-atoms under incidences of 450 nm, 550 nm, and 650 nm, respectively, when the LC distribution is in the 'Switch-Off' state and the x-polarized incidence is converted into y-polarization before hitting the meta-atom. (**f**–**h**) Diagram of transmittance response for different structural parameters of the meta-atoms under incidences of 450 nm, 550 nm, and 650 nm, respectively, when the LC distribution is in the 'Switch-On' state and the x-polarized incidence is maintained before hitting the meta-atom.

The anisotropy characteristics and other optical behaviors in these meta-atoms can be described with the Jones matrix, denoted as $J_{meta} = [t_L, 0; 0, t_S]$, where t_L is the longer optical axis and t_S is the shorter optical axis. Thus, the complex transmittance E_{out} from the meta-atom under an arbitrary incidence E_{in} can be represented by

$$E_{out} = J_{meta} E_{in}. \tag{1}$$

For x-polarized incidence with $E_{in} = [E_{xin}; E_{yin}] = [1; 0]$, E_{out} should be

$$\begin{bmatrix} E_{xout} \\ E_{yout} \end{bmatrix} = \begin{bmatrix} t_L & 0 \\ 0 & t_S \end{bmatrix} \begin{bmatrix} 1 \\ 0 \end{bmatrix} = \begin{bmatrix} t_L \\ 0 \end{bmatrix}, \tag{2}$$

and for y-polarized incidence with $E_{in} = [E_{xin}; E_{yin}] = [0; 1]$, E_{out} should be

$$\begin{bmatrix} E_{xout} \\ E_{yout} \end{bmatrix} = \begin{bmatrix} t_L & 0 \\ 0 & t_S \end{bmatrix} \begin{bmatrix} 0 \\ 1 \end{bmatrix} = \begin{bmatrix} 0 \\ t_S \end{bmatrix}. \tag{3}$$

The transmittance spectra of the proposed dual-channel meta-atoms were simulated using the finite difference time-domain (FDTD) method, specifically employing Lumerical FDTD Solutions. The meta-atoms had a period of 345 nm and a height of 600 nm. To account for fabrication constraints and period sizes, the meta-atoms were subjected to minimum and maximum size constraints of 50 nm and 300 nm, respectively. In order to facilitate the interpretation of the LC molecule behavior, the reorientation process of the LC was simulated accordingly. The equilibrium distribution of the LC director was calculated, and the resulting dielectric tensor field was incorporated into the FDTD simulations to determine the optical response. The computational domain was bounded above and below by a perfectly matched layer (PML) boundary condition with periodic conditions along the x and y directions.

A total of more than 2500 distinct meta-atoms were generated, out of which 64 types were carefully selected for spectral encoding purposes. These 64 structures were arranged in an 8 × 8 configuration, resembling a micro-spectrometer, and were subsequently replicated periodically to form a larger metasurface layer with a 256 × 256 array. When it comes to designing binary-coded apertures using only 0 and 1, an effective design rule that minimizes coherence involves the following considerations [68]: (1) maximizing the separation between one-valued entries within the same row of the coded aperture, (2) reducing the occurrence of vertical clusters in the vertical direction of the coded aperture, and (3) employing a complementary set of codes to minimize correlations, given that each measurement snapshot employs a distinct coded aperture. However, due to the meta-atoms' ability to provide continuous transmittance control across the entire bandwidth, a slightly different yet similar strategy is adopted by employing a genetic algorithm. In both the horizontal and vertical directions, it is crucial to ensure a significant difference in transmittance between each pixel and its neighboring pixels. Furthermore, there should be substantial variation in transmittance for the meta-atoms within the same pixel when subjected to 2-frame snapshots of orthogonally polarized incidences. These requirements are essential to achieve the desired performance and characteristics of the metasurface.

2.2. Dual-Channel Compressive Hyperspectral Imaging with MD-SCSI

The spectral data cube of the incidence can be represented by $f(x, y, \lambda)$, where (x, y) is the position on the metasurface plane and λ is the wavelength. Since every metasurface unit consists of 10 × 10 identical meta-atoms, its transmission spectra are the same as those of a single corresponding meta-atom. The transmission spectra tuning function can be denoted as $T_{ix}(x, y, \lambda)$ and $T_{iy}(x, y, \lambda)$ for x-polarization and y-polarization channels, respectively. Here, $i = 1, 2, 3 \ldots 64$ represents the area number of the metasurface unit within the micro-spectrometer, as illustrated in Figure 4. Consequently, the pre-captured tuned data cube can be represented as $f(x, y, \lambda)T_{ix}(x, y, \lambda)$ and $f(x, y, \lambda)T_{iy}(x, y, \lambda)$. Furthermore, considering $\Omega(\lambda)$ as the spectral response function of the image sensor pixel, the finally detected measurement for pixel (m, n) can be expressed as follows:

$$\begin{cases} g_{mnx} = \int_{(n-1)\Delta}^{n\Delta} \int_{(m-1)\Delta}^{m\Delta} \int_\lambda \Omega(\lambda) f(x,y,\lambda) T_{ix}(x,y,\lambda) \mathrm{d}\lambda \mathrm{d}x \mathrm{d}y \\ g_{mny} = \int_{(n-1)\Delta}^{n\Delta} \int_{(m-1)\Delta}^{m\Delta} \int_\lambda \Omega(\lambda) f(x,y,\lambda) T_{iy}(x,y,\lambda) \mathrm{d}\lambda \mathrm{d}x \mathrm{d}y \end{cases}, \tag{4}$$

where $\Delta = 3.45$ μm is the period of one metasurface unit or one pixel of the image sensor, and (m, n) is the location of a particular pixel and corresponds to the metasurface unit number i from 64 types. If λ is a discrete parameter indicated by s, then the measurement of the pixel (m, n) would be

$$\begin{cases} g_{mnx} = \sum_{s=1}^{N} \Omega_s f_{mns} T_{sx}^i \\ g_{mny} = \sum_{s=1}^{N} \Omega_s f_{mns} T_{sy}^i \end{cases}, \quad (5)$$

where N is the quantity of the spectral bands, and Ω_s is the spectral response of the sensor for the sth spectral band. Supposing that the detected image has P rows and Q columns, add the total g_{mn} and f_{mns} to the detected 2D image $\mathbf{G}$ ($\mathbf{G} \in \mathbb{R}^{P \times Q}$) and the total 3D hyperspectral data cube $\mathbf{F}$ ($\mathbf{F} \in \mathbb{R}^{P \times Q \times S}$), respectively, and denote $\mathbf{g}$ ($\mathbf{g} \in \mathbb{R}^{2PQ \times 1}$) and $\mathbf{f}$ ($\mathbf{f} \in \mathbb{R}^{PQS \times 1}$) as the vectorization of $\mathbf{G}$ and $\mathbf{F}$, respectively. Then the total detected signal can be expressed as

$$\mathbf{g} = \mathbf{Hf}, \quad (6)$$

where $\mathbf{H} = [\mathbf{H}_x; \mathbf{H}_y]$ ($\mathbf{H} \in \mathbb{R}^{2PQ \times PQS}$) is the observation matrix of MD-SCSI with dual polarization channels, as the number 2 in the superscripts of $\mathbf{G}$ and $\mathbf{H}$ denotes both channels. The total detected signal $\mathbf{g}$ can be represented by $\mathbf{g} = [\mathbf{g}_x; \mathbf{g}_y]$, where $\mathbf{g}_x = \mathbf{H}_x \mathbf{f}$ and $\mathbf{g}_y = \mathbf{H}_y \mathbf{f}$.

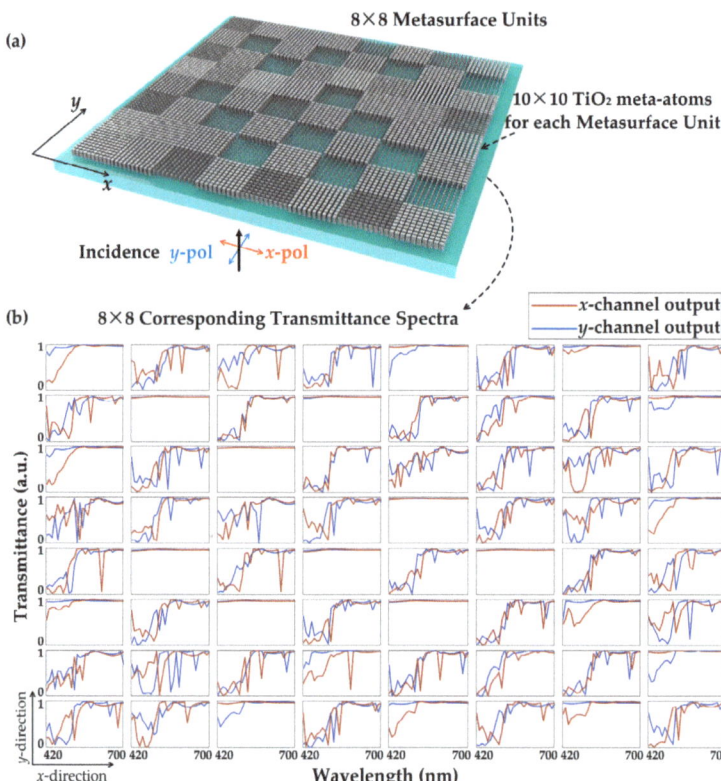

Figure 4. (**a**) View of the micro-spectrometer. Each micro-spectrometer consists of 8 × 8 metasurface units, and each metasurface unit is composed of 10 × 10 TiO$_2$ meta-atoms. (**b**) Transmittance spectra of x-channel (red-line) and y-channel (blue-line) for the selected 8 × 8 metasurface units that make up the micro-spectrometer. The subfigures are arranged according to the relative location distributions of the 64 types of metasurface units along the x–y plane.

2.3. Design of ResNet for Data Cube Reconstruction

For data cube reconstruction, an end-to-end residual neural network (ResNet) is utilized. As a proof of principle, a subset of 29 wavelengths, ranging from 420 to 700 nm with a step size of 10 nm, is selected. The network architecture, depicted in Figure 5, initiates by applying a convolutional layer to transform the input measurements, which have a size of 256 × 256 × 1, into feature maps with dimensions 256 × 256 × K. Subsequently, N residual blocks are incorporated into the network. Finally, another convolutional layer is employed to convert the 256 × 256 × K feature maps into output channels with dimensions 256 × 256 × 29, corresponding to the selected wavelengths. Each residual block, as illustrated in the subfigure of Figure 5, comprises two convolutional layers and is enhanced with a rectified linear unit (ReLU) activation function. This composition enhances its ability to learn and capture the complex relationships within the data.

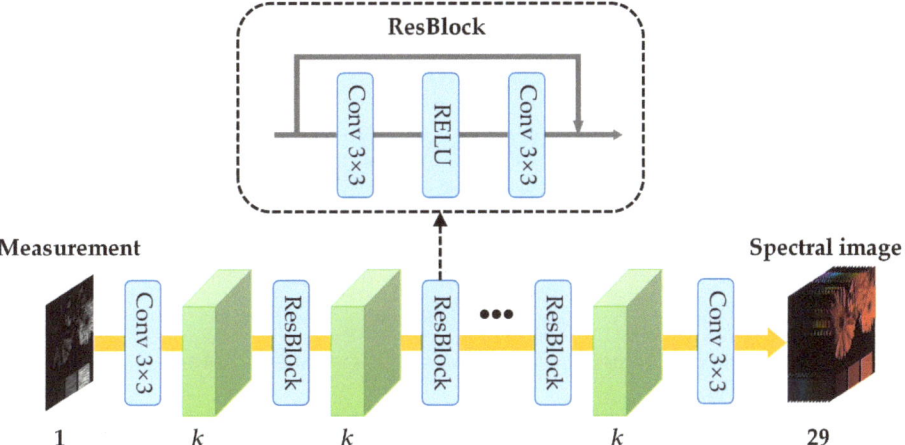

Figure 5. The framework of the utilized E2E ResNet for data cube reconstruction.

In this study, the CAVE dataset, comprising a collection of 32 hyperspectral images with dimensions of 512 × 512 × 32, is utilized for training purposes. Specifically, the parameters K and N are set to 72 and 20, respectively, to optimize the performance of the network. To enhance the training process, data augmentation techniques and spectral interpolation methods are employed, generating an extended dataset consisting of 272 hyperspectral images with dimensions of 1024 × 1024 × 29. For the purpose of testing, a set of 15 scenes obtained from KAIST is selected, each with dimensions of 256 × 256 × 29. These scenes provide a diverse range of real-world scenarios for evaluating the proposed approach. The computational resources employed in this study include an NVIDIA GTX 4090 GPU and an AMD 7950X CPU. These resources contribute to the efficient execution of the training and testing processes, enabling timely and accurate analysis of the hyperspectral data.

The formulated loss function encompasses three distinct components. The first component corresponds to the mean squared error (MSE) loss, quantifying the average squared discrepancy between the predicted values and the ground truth values:

$$L_{\text{mse}} = \left\| \mathbf{F}_{\text{pred}} - \mathbf{F} \right\|_2^2. \tag{7}$$

The second component incorporates the structural similarity (SSIM) loss function, which assesses the structural similarity between the predicted values and the ground truth values by comparing the local patterns, luminance, and contrast of the predicted values with the true values. This loss function facilitates the evaluation of the perceptual quality of the reconstructed data:

$$L_{\text{SSIM}} = SSIM(\mathbf{F}_{\text{pred}}, \mathbf{F}). \tag{8}$$

Lastly, the third component aims to minimize the spectral angle mapping (SAM). The SAM metric determines the similarity between the reconstructed spectra and the true spectra, focusing on the angular difference between them. The goal is to minimize this discrepancy, enhancing the spectral fidelity of the reconstructed data:

$$\text{SAM} = \sum_{i=1}^{256}\sum_{j=1}^{256} \cos^{-1}\left(\frac{\sum_{k=1}^{29} \mathbf{F}_{\text{pred}\,i,j,k}\mathbf{F}_{i,j,k}}{\left(\sum_{k=1}^{29}\mathbf{F}_{\text{pred}\,i,j,k}^2\right)^{1/2}\left(\sum_{k=1}^{29}\mathbf{F}_{i,j,k}^2\right)^{1/2}}\right). \tag{9}$$

Since there is no need to calculate $\cos^{-1}$ while training, to save time, the loss function of SAM is set as:

$$L_{\text{SAM}} = \sum_{i=1}^{256}\sum_{j=1}^{256}\left(\frac{\sum_{k=1}^{29}\mathbf{F}_{\text{pred}\,i,j,k}\mathbf{F}_{i,j,k}}{\left(\sum_{k=1}^{29}\mathbf{F}_{\text{pred}\,i,j,k}^2\right)^{1/2}\left(\sum_{k=1}^{29}\mathbf{F}_{i,j,k}^2\right)^{1/2} + \varepsilon}\right), \tag{10}$$

where $\varepsilon = 1 \times 10^{-9}$ represents a very small value to prevent division by zero. The overall loss function is the weighted sum of these three components:

$$L = L_{\text{mse}} + \alpha_1 L_{\text{SSIM}} - \alpha_2 L_{\text{SAM}}, \tag{11}$$

where $\alpha_1 = 0.05$ and $\alpha_2 = 0.01$ are parameters to balance these three terms for this work.

3. Results

To assess the efficacy of the proposed method, a comprehensive evaluation is conducted utilizing three quantitative metrics to gauge the quality of the reconstructed hyperspectral images. These metrics encompass the SAM, peak signal-to-noise ratio (PSNR), and SSIM. The SAM metric serves as an indicator of spectral accuracy, ranging from 0 to 1. A smaller SAM value signifies a more precise representation of the spectrum. PSNR assesses the overall reconstruction quality, and higher PSNR values indicate superior reconstruction outcomes. SSIM measures the structural similarity between the restored images and the original ones with values ranging from 0 to 1. Larger SSIM values represent minimal distortion in the reconstructed images, implying a higher degree of similarity to the originals. To facilitate a fair and straightforward comparison, the selected test images are scaled from 0 to 1. This normalization simplifies the comparative analysis among the different methods. In the subsequent analysis, the proposed MD-SCSI method is benchmarked against various existing methods for CSI. These methods include CASSI-DBR, CASSI-Net, Meta-DBR, as well as x-pol or y-pol as introduced previously. By comparing the performance of the MD-SCSI method against these alternative approaches, a comprehensive evaluation of its effectiveness is conducted, illustrating its advantages and potential in terms of the quality and speed of spectral reconstructions.

Table 1 showcases the reconstruction outcomes achieved through the application of diverse methodologies to a set of 15 images from the KAIST dataset. The findings firmly substantiate the exceptional reconstruction quality of the proposed MD-SCSI in comparison to the other methods. Across all 15 test images (Figure 6), the MD-SCSI method exhibits notable superiority in terms of SAM reduction, as well as significant enhancements in both PSNR and SSIM measures. These results underscore the remarkable efficacy of the proposed method, affirming its potential to outperform existing techniques in the domain of HSI reconstruction.

Table 1. SAM, PSNR, and SSIM comparison of different methods for 15 images in the KAIST dataset.

Image Number	Quality Metrics	CASSI-DBR	CASSI-Net	Meta-TwIST	x-pol	y-pol	MD-SCSI
Image 1	SAM	0.0860	0.1287	0.1095	0.0972	0.0989	**0.0570**
	PSNR	30.80	27.97	29.15	30.81	30.44	**35.59**
	SSIM	0.8867	0.8969	0.8713	0.9370	0.9357	**0.9643**
Image 2	SAM	0.0904	0.1364	0.1692	0.1220	0.1098	**0.0714**
	PSNR	35.03	31.65	32.92	35.51	35.61	**39.93**
	SSIM	0.9472	0.9393	0.9279	0.9739	0.9741	**0.9891**
Image 3	SAM	0.1332	0.1675	0.2689	0.1754	0.1604	**0.1106**
	PSNR	35.48	36.38	33.77	38.35	38.10	**41.84**
	SSIM	0.9533	0.9747	0.9275	0.9785	0.9808	**0.9914**
Image 4	SAM	0.1974	0.1314	0.2377	0.1139	0.1225	**0.0751**
	PSNR	30.28	31.57	28.94	33.01	31.96	**38.35**
	SSIM	0.8795	0.9415	0.8541	0.9533	0.9470	**0.9830**
Image 5	SAM	0.1036	0.1385	0.1324	0.1238	0.1266	**0.0765**
	PSNR	36.80	34.99	35.63	36.60	37.12	**40.95**
	SSIM	0.9557	0.9546	0.9508	0.9669	0.9672	**0.9845**
Image 6	SAM	0.0993	0.1105	0.1636	0.1078	0.1033	**0.0686**
	PSNR	35.17	34.44	33.65	36.20	35.55	**40.64**
	SSIM	0.9626	0.9728	0.9499	0.9820	0.9820	**0.9916**
Image 7	SAM	0.1001	0.1208	0.1595	0.1123	0.1065	**0.0692**
	PSNR	30.48	29.84	29.33	32.43	32.40	**36.89**
	SSIM	0.9166	0.9340	0.9005	0.9524	0.9527	**0.9738**
Image 8	SAM	0.1194	0.1378	0.1970	0.1476	0.1494	**0.0752**
	PSNR	35.14	32.21	32.73	32.58	32.58	**39.21**
	SSIM	0.9368	0.9470	0.9136	0.9570	0.9561	**0.9856**
Image 9	SAM	0.0799	0.0913	0.0946	0.0642	0.0700	**0.0449**
	PSNR	36.70	33.76	35.68	37.49	36.41	**39.78**
	SSIM	0.9388	0.9455	0.9402	0.9705	0.9661	**0.9781**
Image 10	SAM	0.0813	0.1081	0.1518	0.1016	0.0870	**0.0580**
	PSNR	42.37	40.26	39.45	41.41	42.09	**47.90**
	SSIM	0.9820	0.9827	0.9701	0.9894	0.9906	**0.9959**
Image 11	SAM	0.0967	0.1427	0.2434	0.1310	0.1248	**0.0893**
	PSNR	39.2121	37.70	37.12	40.57	40.76	**43.70**
	SSIM	0.9656	0.9743	0.9335	0.9841	0.9870	**0.9929**
Image 12	SAM	0.2171	0.1864	0.2920	0.1639	0.1557	**0.1037**
	PSNR	35.4471	36.3229	33.4164	39.5796	39.9063	**44.5338**
	SSIM	0.9060	0.9633	0.8706	0.9790	0.9800	**0.9917**
Image 13	SAM	0.1030	0.1208	0.1945	0.1202	0.1128	**0.0785**
	PSNR	38.1660	38.4325	35.0692	40.8228	40.5658	**44.1873**
	SSIM	0.9565	0.9777	0.9321	0.9855	0.9863	**0.9926**
Image 14	SAM	0.1071	0.1440	0.2363	0.1419	0.1304	**0.0878**
	PSNR	33.6037	33.3215	32.2650	36.1854	35.8398	**40.6773**
	SSIM	0.9641	0.9751	0.9454	0.9817	0.9832	**0.9921**
Image 15	SAM	0.0897	0.1172	0.1928	0.1253	0.1124	**0.0798**
	PSNR	38.8602	39.2061	37.3496	41.3150	41.3412	**44.2754**
	SSIM	0.9751	0.9821	0.9608	0.9861	0.9874	**0.9937**
Average	SAM	0.1136	0.1321	0.1896	0.1232	0.1180	**0.0764**
	PSNR	35.5685	34.5362	33.7651	36.8566	36.7112	**41.2303**
	SSIM	0.9418	0.9574	0.9232	0.9718	0.9718	**0.9867**

Figure 6. 15 testing images from the KAIST dataset for simulation.

Moreover, an analysis of the comprehensive visual comparison results along with detailed information would provide further substantiation of the effectiveness of the proposed method. These illustrative demonstrations prove the improved performance and advantages offered by the proposed MD-SCSI when compared to alternative methods. By combining quantitative metrics with visual analysis, a compelling demonstration can be achieved, consistently showcasing the superior reconstruction quality achieved by the proposed method in comparison to other methods. This comprehensive evaluation, including both objective metrics and qualitative visual assessment, reinforces the efficacy of the proposed method and establishes its superiority in HSI reconstruction.

Figure 7 presents the reconstruction results of the different methods on three selected test images, and the reconstructed results produced with MD-SCSI (a7, b7, and c7) keep spatial details well with fine color fidelity, thereby demonstrating their superior reconstruction quality and precision. Conversely, it can be observed that CASSI-DBR exhibits significant noise, leading to noticeable blur, color distortion, and colored speckle noise as could be viewed in the zoomed-in versions. CASSI-Net and Meta-DBR achieve clearer reconstructed results; however, they still suffer from substantial color distortion or colored speckle noise overall. x-pol and y-pol produce satisfactory results with significantly improved clarity and colors that are closer to the ground truth. Nonetheless, some color distortion still persists, particularly in colors such as yellow and red. Finally, MD-SCSI yields the visually best results, characterized by clear outputs and colors that closely resemble the original images.

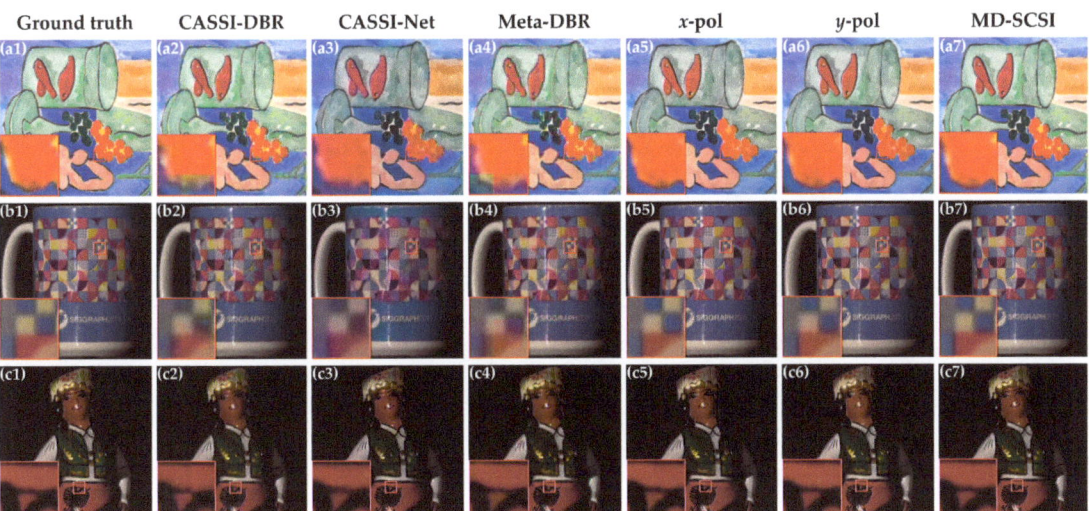

Figure 7. Visual comparison of reconstruction results using different methods with the 3D spectral data being synthesized into RGB images for intuitive display. (**a1–a7**) Ground truth compared with the results of CASSI-DBR, CASSI-Net, Meta-DBR, x-pol, y-pol, and MD-SCSI for image 1 in the KAIST dataset. (**b1–b7**) Ground truth compared with the results of CASSI-DBR, CASSI-Net, Meta-DBR, x-pol, y-pol, and MD-SCSI for image 2 in the KAIST dataset. (**c1–c7**) Ground truth compared with the results of CASSI-DBR, CASSI-Net, Meta-DBR, x-pol, y-pol, and MD-SCSI for image 3 in the KAIST dataset.

In order to facilitate a more intuitive comparison in the spectral domain, Figure 8 illustrates the recovered spectra from two randomly chosen spatial positions in test images 4 and 5. It can be observed from Figure 8 that the spectral signatures reconstructed with MD-SCSI (orange lines) exhibit the best similarity to the ground truth spectra (yellow lines), and the average mean square errors (MSE) of the reconstructed spectra shown in Table 2 demonstrate that MD-SCSI is the most accurate statistically, further proving that MD-SCSI outperforms the other methods.

Table 2. Average mean square error of the spectra for the selected points depicted in Figure 8.

Image	CASSI-DBR	CASSI-Net	Meta-DBR	x-pol	y-pol	MD-SCSI
Image 4	1.771×10^{-4}	1.800×10^{-4}	3.565×10^{-4}	1.565×10^{-4}	2.425×10^{-4}	$\mathbf{9.184 \times 10^{-5}}$
Image 5	9.310×10^{-5}	1.109×10^{-4}	1.645×10^{-4}	1.671×10^{-4}	1.456×10^{-4}	$\mathbf{3.119 \times 10^{-5}}$

Figure 9 displays the reconstruction results of two test images, image 6 and image 7, in three specific spectral channels of 450 nm, 550 nm, and 650 nm. It can be observed that the reconstruction results of CASSI-DBR, CASSI-Net, and Meta-DBR introduce obvious noise, resulting in relatively blurred details, shape deformations, and erroneous details in the 650 nm channel. The reconstruction results of x-pol and y-pol are closer to the ground truth; however, there are still incorrect details in the 650 nm band. Finally, MD-SCSI provides the most faithful representation of the ground truth, as it closely resembles the original images, exhibits fewer undesirable visual artifacts, and provides visually better reconstructions than the other methods without clear noises or erroneous details in the 650 nm band. This validates the superiority of MD-SCSI in this work.

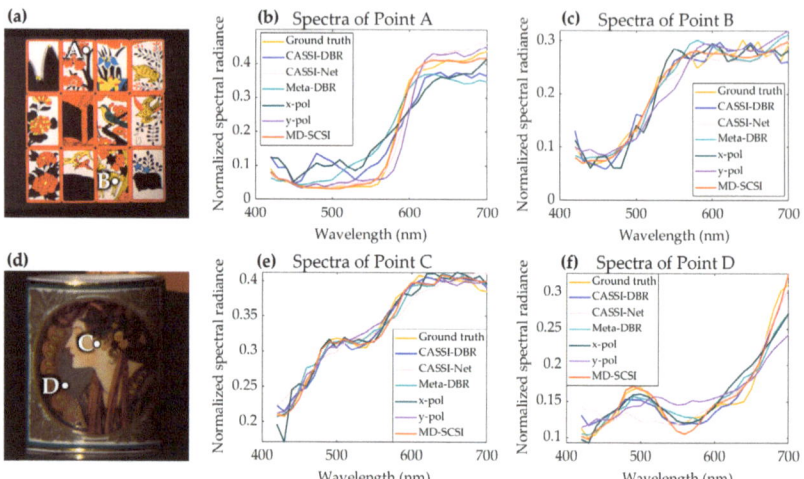

Figure 8. (**a**) Image 4 from the KAIST dataset with two randomly selected points, A and B, with the spectra of A and B that are reconstructed with different methods and depicted in (**b**,**c**), respectively. (**d**) Image 5 from the KAIST dataset with two randomly selected points, C and D, with the spectra of C and D that are reconstructed with different methods and depicted in (**e**,**f**), respectively.

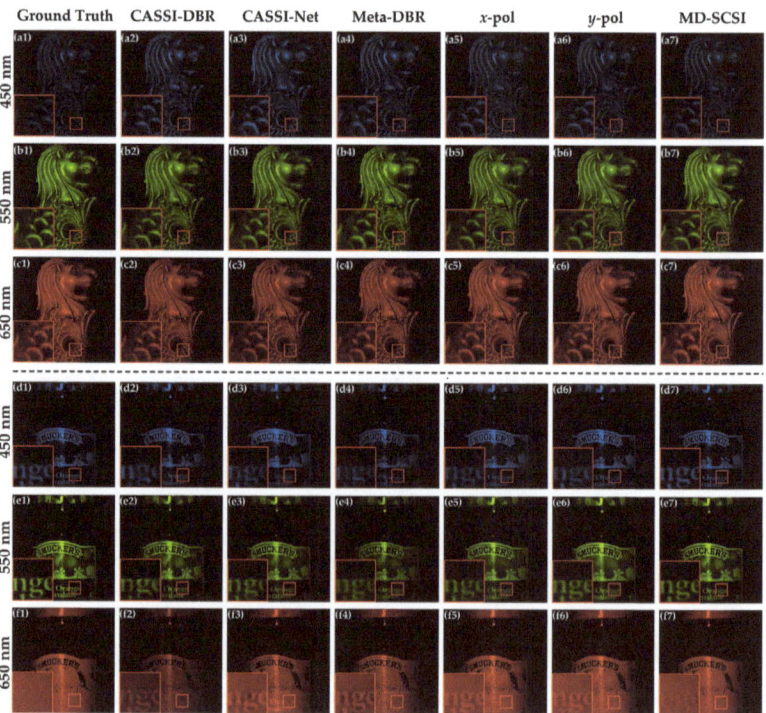

Figure 9. Visual comparison of image 6 for (**a1**–**a7**) 450 nm, (**b1**–**b7**) 550 nm, and (**c1**–**c7**) 650 nm with various methods compared to the corresponding ground truth, and visual comparison of image 7 for (**d1**–**d7**) 450 nm, (**e1**–**e7**) 550 nm, and (**f1**–**f7**) 650 nm with various methods compared to the corresponding ground truth.

Furthermore, Figure 10 showcases a comparison between the entire reconstructed hyperspectral image generated with MD-SCSI and the original image 8, and it clearly demonstrates that MD-SCSI generates reconstructed results that are visually pleasing and very close to the ground truths in terms of spatial structure, color fidelity, and edge details without noticeable noises or distortions.

Figure 10. The ground truth of full-band spectra of image 8 with its reconstructed results of full-band spectra with the proposed MD-SCSI method.

Additionally, the average running time of MD-SCSI with the other contrasting methods on the 15 test images is listed in Table 3. Generally, the reconstructions of 15 test images are repeated 10 times for all six methods, including the proposed method, MD-SCSI, as well as the other contrasting methods. Thus, the running time for each method is the average of its corresponding 150 reconstruction times, respectively. Specifically, the PyTorch profiler tool is utilized for the evaluation of deep-learning-based methods (MD-SCSI, x-pol, y-pol, and CASSI-Net), and the timing function of MATLAB is adopted for the evaluation of the dictionary learning-based methods (CASSI-DBR and Meta-DBR). It could be noted that the methods based on neural networks require significantly less processing time, approximately in the order of 1×10^{-3}, compared to the methods using iterative CS algorithms and dictionary learning, and the proposed E2E ResNet algorithm, which is optimized for MD-SCSI, performs the fastest reconstruction speed when similarly applied to x-pol, y-pol, and MD-SCSI. The deep-learning-based methods excel in rapid reconstruction, and this is because of the relatively straightforward matrix operations from parallel processing on GPUs as well as the inherent simplicity of the reconstruction process.

Table 3. Average running time on the test dataset for different methods.

Method	CASSI-DBR	CASSI-Net	Meta-DBR	x-pol	y-pol	MD-SCSI
Running Time	138.1021 s	118.3007 ms	135.5728 s	115.3872 ms	116.9829 ms	115.9519 ms

4. Conclusions

In conclusion, this paper proposes MD-SCSI, a compact snapshot hyperspectral compressive imaging system that leverages dual-channel switchable metasurface filters in conjunction with a deep-learning-empowered reconstruction algorithm based on compressive sensing theory. MD-SCSI presents a compact framework for snapshot hyperspectral imaging, eliminating the need for dispersion elements or additional cameras. This pioneering approach capitalizes on the unique advantages offered by dual-channel metasurfaces as SCSI hardware encoders while employing an end-to-end residual neural network for HSI reconstruction, thus demonstrating superiority in both system compactness and reconstruction performance. In detail, the dual-channel switchable metasurface filters are arranged according to minimization of coherence and integrated by TNLCs with broadband linear polarization conversion abilities. These metasurfaces enable rapid conversion between different voltages, generating 2-frame snapshots of spectral information with distinct transmittance spectra for each independent input channel, facilitating the reconstruction of spectral data cubes. Furthermore, the specially optimized end-to-end ResNet, tailored to the characteristics of the dual-channel metasurface filters, enables efficient HSI reconstruction by processing the 2-frame snapshot measurements acquired from orthogonal polarization channels of MD-SCSI. This reconstruction process achieves exceptional quality in terms of structural fidelity, color fidelity, spatial resolution, and speed. The comparison of the HSI reconstruction results obtained with MD-SCSI with the other approaches demonstrates the effectiveness, accuracy, as well as speediness of MD-SCSI, which greatly outperforms the other HSI reconstruction methods. Consequently, MD-SCSI exhibits tremendous potential for compact, accurate, and rapid SCSI and could be applied across various domains, including but not limited to food safety, biomedical imaging, precision agriculture, and object detection, among others.

Author Contributions: Conceptualization, C.W. and X.L.; Data curation, C.W. and X.L.; Formal analysis, C.W., X.L., Y.Z., Y.S. and Z.Y.; Funding acquisition, Z.Z.; Investigation, C.W. and X.L.; Methodology, C.W. and X.L.; Project administration, Z.Z.; Resources, Z.Z.; Software, C.W. and X.L.; Supervision, Z.Z.; Validation, C.W. and X.L.; Visualization, C.W. and X.L.; Writing—original draft, C.W. and X.L.; Writing—review and editing, C.W., X.L., Y.Z., Y.S. and Z.Y. All authors have read and agreed to the published version of the manuscript.

Funding: This research was funded by the National Key Research and Development Program of China, grant number 2022YFF0705500.

Data Availability Statement: The data underlying the results presented in this paper are not publicly available at this time but may be obtained from the authors upon reasonable request.

Conflicts of Interest: The authors declare no conflict of interest.

References

1. Makki, I.; Younes, R.; Francis, C.; Bianchi, T.; Zucchetti, M. A Survey of Landmine Detection Using Hyperspectral Imaging. *ISPRS J. Photogramm. Remote Sens.* **2017**, *124*, 40–53. [CrossRef]
2. Meng, Z.; Qiao, M.; Ma, J.; Yu, Z.; Xu, K.; Yuan, X. Snapshot Multispectral Endomicroscopy. *Opt. Lett.* **2020**, *45*, 3897–3900. [CrossRef]
3. Qin, J.; Chao, K.; Kim, M.S.; Lu, R.; Burks, T.F. Hyperspectral and Multispectral Imaging for Evaluating Food Safety and Quality. *J. Food Eng.* **2013**, *118*, 157–171. [CrossRef]
4. Mahesh, S.; Jayas, D.S.; Paliwal, J.; White, N.D.G. Hyperspectral Imaging to Classify and Monitor Quality of Agricultural Materials. *J. Stored Prod. Res.* **2015**, *61*, 17–26. [CrossRef]

5. Plaza, A.; Benediktsson, J.A.; Boardman, J.W.; Brazile, J.; Bruzzone, L.; Camps-Valls, G.; Chanussot, J.; Fauvel, M.; Gamba, P.; Gualtieri, A.; et al. Recent Advances in Techniques for Hyperspectral Image Processing. *Remote Sens. Environ.* **2009**, *113*, S110–S122. [CrossRef]
6. Yuan, X.; Brady, D.J.; Katsaggelos, A.K. Snapshot Compressive Imaging: Theory, Algorithms, and Applications. *IEEE Signal Process. Mag.* **2021**, *38*, 65–88. [CrossRef]
7. Zhang, Q.; Yu, Z.; Liu, X.; Wang, C.; Zheng, Z. End-to-End Joint Optimization of Metasurface and Image Processing for Compact Snapshot Hyperspectral Imaging. *Opt. Commun.* **2023**, *530*, 129154. [CrossRef]
8. Arce, G.R.; Brady, D.J.; Carin, L.; Arguello, H.; Kittle, D.S. Compressive Coded Aperture Spectral Imaging. *IEEE Signal Process. Mag.* **2014**, *31*, 105–115. [CrossRef]
9. Liu, X.; Yu, Z.; Zheng, S.; Li, Y.; Tao, X.; Wu, F.; Xie, Q.; Sun, Y.; Wang, C.; Zheng, Z. Residual Image Recovery Method Based on the Dual-Camera Design of a Compressive Hyperspectral Imaging System. *Opt. Express* **2022**, *30*, 20100. [CrossRef]
10. Sun, M.-J.; Meng, L.-T.; Edgar, M.P.; Padgett, M.J.; Radwell, N. A Russian Dolls Ordering of the Hadamard Basis for Compressive Single-Pixel Imaging. *Sci. Rep.* **2017**, *7*, 3464. [CrossRef]
11. Lin, X.; Wetzstein, G.; Liu, Y.; Dai, Q. Dual-Coded Compressive Hyperspectral Imaging. *Opt. Lett.* **2014**, *39*, 2044–2047. [CrossRef]
12. Donoho, D. Compressed Sensing. *IEEE Trans. Inf. Theory* **2006**, *52*, 1289–1306. [CrossRef]
13. Candes, E.; Romberg, J.; Tao, T. Robust Uncertainty Principles: Exact Signal Reconstruction from Highly Incomplete Frequency Information. *IEEE Trans. Inf. Theory* **2006**, *52*, 489–509. [CrossRef]
14. Gehm, M.E.; John, R.; Brady, D.J.; Willett, R.M.; Schulz, T.J. Single-Shot Compressive Spectral Imaging with a Dual-Disperser Architecture. *Opt. Express* **2007**, *15*, 14013–14027. [CrossRef]
15. Wagadarikar, A.; John, R.; Willett, R.; Brady, D. Single Disperser Design for Coded Aperture Snapshot Spectral Imaging. *Appl. Opt.* **2008**, *47*, B44–B51. [CrossRef]
16. Kittle, D.; Choi, K.; Wagadarikar, A.; Brady, D.J. Multiframe Image Estimation for Coded Aperture Snapshot Spectral Imagers. *Appl. Opt.* **2010**, *49*, 6824–6833. [CrossRef]
17. Wang, L.; Xiong, Z.; Gao, D.; Shi, G.; Wu, F. Dual-Camera Design for Coded Aperture Snapshot Spectral Imaging. *Appl. Opt.* **2015**, *54*, 848–858. [CrossRef]
18. Arguello, H.; Arce, G.R. Colored Coded Aperture Design by Concentration of Measure in Compressive Spectral Imaging. *IEEE Trans. Image* **2014**, *23*, 1896–1908. [CrossRef]
19. Lin, X.; Liu, Y.; Wu, J.; Dai, Q. Spatial-Spectral Encoded Compressive Hyperspectral Imaging. *ACM Trans. Graph.* **2014**, *33*, 1–11. [CrossRef]
20. Huang, L.; Luo, R.; Liu, X.; Hao, X. Spectral Imaging with Deep Learning. *Light Sci. Appl.* **2022**, *11*, 61. [CrossRef]
21. Wang, L.; Wu, Z.; Zhong, Y.; Yuan, X. Snapshot Spectral Compressive Imaging Reconstruction Using Convolution and Contextual Transformer. *Photon. Res.* **2022**, *10*, 1848–1858. [CrossRef]
22. Bioucas-Dias, J.M.; Figueiredo, M.A.T. A New TwIST: Two-Step Iterative Shrinkage/Thresholding Algorithms for Image Restoration. *IEEE Trans. Image Process.* **2007**, *16*, 2992–3004. [CrossRef]
23. Chambolle, A. An Algorithm for Total Variation Minimization and Applications. *J. Math. Imaging Vis.* **2004**, *20*, 89–97.
24. Nguyen, R.M.H.; Prasad, D.K.; Brown, M.S. Training-Based Spectral Reconstruction from a Single RGB Image. In *Computer Vision—ECCV 2014, Proceedings of the 13th European Conference, Zurich, Switzerland, 6–12 September 2014*; Fleet, D., Pajdla, T., Schiele, B., Tuytelaars, T., Eds.; Springer International Publishing: Cham, Switzerland, 2014; pp. 186–201.
25. Arad, B.; Ben-Shahar, O. Sparse Recovery of Hyperspectral Signal from Natural RGB Images. In *Computer Vision—ECCV 2016, Proceedings of the 14th European Conference, Amsterdam, The Netherlands, 11–14 October 2016*; Leibe, B., Matas, J., Sebe, N., Welling, M., Eds.; Springer International Publishing: Cham, Switzerland, 2016; pp. 19–34.
26. Hsiao, H.-H.; Chu, C.H.; Tsai, D.P. Fundamentals and Applications of Metasurfaces. *Small Methods* **2017**, *1*, 1600064. [CrossRef]
27. Sun, S.; He, Q.; Hao, J.; Xiao, S.; Zhou, L. Electromagnetic Metasurfaces: Physics and Applications. *Adv. Opt. Photon.* **2019**, *11*, 380–479. [CrossRef]
28. Genevet, P.; Capasso, F.; Aieta, F.; Khorasaninejad, M.; Devlin, R. Recent Advances in Planar Optics: From Plasmonic to Dielectric Metasurfaces. *Optica* **2017**, *4*, 139–152. [CrossRef]
29. Yu, N.; Genevet, P.; Aieta, F.; Kats, M.A.; Blanchard, R.; Aoust, G.; Tetienne, J.-P.; Gaburro, Z.; Capasso, F. Flat Optics: Controlling Wavefronts With Optical Antenna Metasurfaces. *IEEE J. Sel. Top. Quantum Electron.* **2013**, *19*, 4700423. [CrossRef]
30. Kamali, S.M.; Arbabi, E.; Arbabi, A.; Faraon, A. A Review of Dielectric Optical Metasurfaces for Wavefront Control. *Nanophotonics* **2018**, *7*, 1041–1068. [CrossRef]
31. Rubin, N.A.; Chevalier, P.; Juhl, M.; Tamagnone, M.; Chipman, R.; Capasso, F. Imaging Polarimetry through Metasurface Polarization Gratings. *Opt. Express* **2022**, *30*, 9389. [CrossRef]
32. Hu, Y.; Wang, X.; Luo, X.; Ou, X.; Li, L.; Chen, Y.; Yang, P.; Wang, S.; Duan, H. All-Dielectric Metasurfaces for Polarization Manipulation: Principles and Emerging Applications. *Nanophotonics* **2020**, *9*, 3755–3780. [CrossRef]
33. Wang, C.; Sun, Y.; Zhang, Q.; Yu, Z.; Tao, C.; Zhang, J.; Wu, F.; Wu, R.; Zheng, Z. Continuous-Zoom Bifocal Metalens by Mutual Motion of Cascaded Bilayer Metasurfaces in the Visible. *Opt. Express* **2021**, *29*, 26569. [CrossRef]
34. Wang, C.; Liu, S.; Sun, Y.; Tao, X.; Sun, P.; Zhang, J.; Tao, C.; Wu, R.; Wu, F.; Zheng, Z. Tunable Beam Splitter Using Bilayer Geometric Metasurfaces in the Visible Spectrum. *Opt. Express* **2020**, *28*, 28672. [CrossRef]

35. Arbabi, A.; Horie, Y.; Bagheri, M.; Faraon, A. Dielectric Metasurfaces for Complete Control of Phase and Polarization with Subwavelength Spatial Resolution and High Transmission. *Nat. Nanotechnol.* **2015**, *10*, 937-U190. [CrossRef]
36. Yang, J.; Cui, K.; Cai, X.; Xiong, J.; Zhu, H.; Rao, S.; Xu, S.; Huang, Y.; Liu, F.; Feng, X.; et al. Ultraspectral Imaging Based on Metasurfaces with Freeform Shaped Meta-Atoms. *Laser Photonics Rev.* **2022**, *16*, 2100663. [CrossRef]
37. Tittl, A.; Leitis, A.; Liu, M.; Yesilkoy, F.; Choi, D.-Y.; Neshev, D.N.; Kivshar, Y.S.; Altug, H. Imaging-Based Molecular Barcoding with Pixelated Dielectric Metasurfaces. *Science* **2018**, *360*, aas9768. [CrossRef]
38. Yang, Z.; Albrow-Owen, T.; Cai, W.; Hasan, T. Miniaturization of Optical Spectrometers. *Science* **2021**, *371*, eabe0722. [CrossRef]
39. Julian, M.N.; Williams, C.; Borg, S.; Bartram, S.; Kim, H.J. Reversible Optical Tuning of GeSbTe Phase-Change Metasurface Spectral Filters for Mid-Wave Infrared Imaging. *Optica* **2020**, *7*, 746–754. [CrossRef]
40. Wang, S.-W.; Xia, C.; Chen, X.; Lu, W.; Li, M.; Wang, H.; Zheng, W.; Zhang, T. Concept of a High-Resolution Miniature Spectrometer Using an Integrated Filter Array. *Opt. Lett.* **2007**, *32*, 632–634. [CrossRef]
41. Pervez, N.K.; Cheng, W.; Jia, Z.; Cox, M.P.; Edrees, H.M.; Kymissis, I. Photonic Crystal Spectrometer. *Opt. Express* **2010**, *18*, 8277–8285. [CrossRef]
42. Du, K.; Barkaoui, H.; Zhang, X.; Jin, L.; Song, Q.; Xiao, S. Optical Metasurfaces towards Multifunctionality and Tunability. *Nanophotonics* **2022**, *11*, 1761–1781. [CrossRef]
43. Kim, J.; Seong, J.; Yang, Y.; Moon, S.-W.; Badloe, T.; Rho, J. Tunable Metasurfaces towards Versatile Metalenses and Metaholograms: A Review. *Adv. Photon.* **2022**, *4*, 024001. [CrossRef]
44. Overvig, A.C.; Malek, S.C.; Yu, N. Multifunctional Nonlocal Metasurfaces. *Phys. Rev. Lett.* **2020**, *125*, 017402. [CrossRef] [PubMed]
45. Fan, K.; Averitt, R.D.; Padilla, W.J. Active and Tunable Nanophotonic Metamaterials. *Nanophotonics* **2022**, *11*, 3769–3803. [CrossRef]
46. Kowerdziej, R.; Ferraro, A.; Zografopoulos, D.C.; Caputo, R. Soft-Matter-Based Hybrid and Active Metamaterials. *Adv. Opt. Mater.* **2022**, *10*, 2200750. [CrossRef]
47. Buchnev, O.; Ou, J.Y.; Kaczmarek, M.; Zheludev, N.I.; Fedotov, V.A. Electro-Optical Control in a Plasmonic Metamaterial Hybridised with a Liquid-Crystal Cell. *Opt. Express* **2013**, *21*, 1633–1638. [CrossRef] [PubMed]
48. Decker, M.; Kremers, C.; Minovich, A.; Staude, I.; Miroshnichenko, A.E.; Chigrin, D.; Neshev, D.N.; Jagadish, C.; Kivshar, Y.S. Electro-Optical Switching by Liquid-Crystal Controlled Metasurfaces. *Opt. Express* **2013**, *21*, 8879–8885. [CrossRef] [PubMed]
49. Dolan, J.A.; Cai, H.; Delalande, L.; Li, X.; Martinson, A.B.F.; de Pablo, J.J.; López, D.; Nealey, P.F. Broadband Liquid Crystal Tunable Metasurfaces in the Visible: Liquid Crystal Inhomogeneities Across the Metasurface Parameter Space. *ACS Photonics* **2021**, *8*, 567–575. [CrossRef]
50. Hu, Y.; Ou, X.; Zeng, T.; Lai, J.; Zhang, J.; Li, X.; Luo, X.; Li, L.; Fan, F.; Duan, H. Electrically Tunable Multifunctional Polarization-Dependent Metasurfaces Integrated with Liquid Crystals in the Visible Region. *Nano Lett.* **2021**, *21*, 4554–4562. [CrossRef]
51. Palermo, G.; Lininger, A.; Guglielmelli, A.; Ricciardi, L.; Nicoletta, G.; De Luca, A.; Park, J.-S.; Lim, S.W.D.; Meretska, M.L.; Capasso, F.; et al. All-Optical Tunability of Metalenses Permeated with Liquid Crystals. *ACS Nano* **2022**, *16*, 16539–16548. [CrossRef]
52. Ou, X.; Zeng, T.; Zhang, Y.; Jiang, Y.; Gong, Z.; Fan, F.; Jia, H.; Duan, H.; Hu, Y. Tunable Polarization-Multiplexed Achromatic Dielectric Metalens. *Nano Lett.* **2022**, *22*, 10049–10056. [CrossRef]
53. Choi, I.; Jeon, D.S.; Nam, G.; Gutierrez, D.; Kim, M.H. High-Quality Hyperspectral Reconstruction Using a Spectral Prior. *ACM Trans. Graph.* **2017**, *36*, 1–13. [CrossRef]
54. SJeon, D.S.; Baek, S.-H.; Yi, S.; Fu, Q.; Dun, X.; Heidrich, W.; Kim, M.H. Compact Snapshot Hyperspectral Imaging with Diffracted Rotation. *ACM Trans. Graph. (TOG)* **2019**, *38*, 1–13. [CrossRef]
55. Hauser, J.; Zeligman, A.; Averbuch, A.; Zheludev, V.A.; Nathan, M. DD-Net: Spectral Imaging from a Monochromatic Dispersed and Diffused Snapshot. *Appl. Opt.* **2020**, *59*, 11196–11208. [CrossRef]
56. Xiong, Z.; Shi, Z.; Li, H.; Wang, L.; Liu, D.; Wu, F. HSCNN: CNN-Based Hyperspectral Image Recovery from Spectrally Undersampled Projections. In Proceedings of the 2017 IEEE International Conference on Computer Vision Workshops (ICCVW), Venice, Italy, 22–29 October 2017; pp. 518–525.
57. Zhang, T.; Fu, Y.; Wang, L.; Huang, H. Hyperspectral Image Reconstruction Using Deep External and Internal Learning. In Proceedings of the 2019 IEEE/CVF International Conference on Computer Vision (ICCV), Seoul, Republic of Korea, 27 October–2 November 2019; pp. 8558–8567.
58. Arad, B.; Timofte, R.; Ben-Shahar, O.; Lin, Y.-T.; Finlayson, G.; Givati, S.; Li, J.; Wu, C.; Song, R.; Li, Y.; et al. NTIRE 2020 Challenge on Spectral Reconstruction from an RGB Image. In Proceedings of the 2020 IEEE/CVF Conference on Computer Vision and Pattern Recognition Workshops (CVPRW), Seattle, WA, USA, 14–19 June 2020; pp. 1806–1822.
59. Miao, X.; Yuan, X.; Pu, Y.; Athitsos, V. Lambda-Net: Reconstruct Hyperspectral Images From a Snapshot Measurement. In Proceedings of the 2019 IEEE/CVF International Conference on Computer Vision (ICCV), Seoul, Republic of Korea, 27 October–2 November 2019; pp. 4058–4068.
60. Vaswani, A.; Shazeer, N.; Parmar, N.; Uszkoreit, J.; Jones, L.; Gomez, A.N.; Kaiser, Ł.; Polosukhin, I. Attention Is All You Need. In Proceedings of the 31st International Conference on Neural Information Processing Systems, Long Beach, CA, USA, 4 December 2017; Curran Associates Inc.: Red Hook, NY, USA, 2017; pp. 6000–6010.
61. He, K.; Zhang, X.; Ren, S.; Sun, J. Deep Residual Learning for Image Recognition. In Proceedings of the 2016 IEEE Conference on Computer Vision and Pattern Recognition (CVPR), Las Vegas, NV, USA, 27–30 June 2016; pp. 770–778.

62. Kohei, Y.; Han, X.-H. Deep Residual Attention Network for Hyperspectral Image Reconstruction. In Proceedings of the 2020 25th International Conference on Pattern Recognition (ICPR), Milan, Italy, 10–15 January 2021; pp. 8547–8553.
63. Siefke, T.; Kroker, S.; Pfeiffer, K.; Puffky, O.; Dietrich, K.; Franta, D.; Ohlídal, I.; Szeghalmi, A.; Kley, E.-B.; Tünnermann, A. Materials Pushing the Application Limits of Wire Grid Polarizers Further into the Deep Ultraviolet Spectral Range. *Adv. Opt. Mater.* **2016**, *4*, 1780–1786. [CrossRef]
64. Arosa, Y.; de la Fuente, R. Refractive Index Spectroscopy and Material Dispersion in Fused Silica Glass. *Opt. Lett.* **2020**, *45*, 4268–4271. [CrossRef] [PubMed]
65. Li, N.; Xu, Z.; Dong, Y.; Hu, T.; Zhong, Q.; Fu, Y.H.; Zhu, S.; Singh, N. Large-Area Metasurface on CMOS-Compatible Fabrication Platform: Driving Flat Optics from Lab to Fab. *Nanophotonics* **2020**, *9*, 3071–3087. [CrossRef]
66. Wu, S.-T.; Wu, C.; Warenghem, W.; Ismaili, M. Refractive Index Dispersions of Liquid Crystals. In *Display Technologies*; SPIE: Bellingham, WA, USA, 1992; Volume 1815, pp. 179–187.
67. König, T.A.F.; Ledin, P.A.; Kerszulis, J.; Mahmoud, A.; El-Sayed, M.A.; Reynolds, J.R.; Tsukruk, V.V. Electrically Tunable Plasmonic Behavior of Nanocube–Polymer Nanomaterials Induced by a Redox-Active Electrochromic Polymer. *ACS Nano* **2014**, *8*, 6182–6192. [CrossRef] [PubMed]
68. Correa, C.V.; Arguello, H.; Arce, G.R. Spatiotemporal Blue Noise Coded Aperture Design for Multi-Shot Compressive Spectral Imaging. *J. Opt. Soc. Am. A* **2016**, *33*, 2312–2322. [CrossRef]

Disclaimer/Publisher's Note: The statements, opinions and data contained in all publications are solely those of the individual author(s) and contributor(s) and not of MDPI and/or the editor(s). MDPI and/or the editor(s) disclaim responsibility for any injury to people or property resulting from any ideas, methods, instructions or products referred to in the content.

Article

Optical, Photophysical, and Electroemission Characterization of Blue Emissive Polymers as Active Layer for OLEDs

Despoina Tselekidou [1,*], Kyparisis Papadopoulos [1], Konstantinos C. Andrikopoulos [2], Aikaterini K. Andreopoulou [2], Joannis K. Kallitsis [2], Stergios Logothetidis [1,3], Argiris Laskarakis [1] and Maria Gioti [1,*]

[1] Nanotechnology Laboratory LTFN, Department of Physics, Aristotle University of Thessaloniki, 54124 Thessaloniki, Greece; kypapado@physics.auth.gr (K.P.); logot@auth.gr (S.L.); alask@physics.auth.gr (A.L.)
[2] Department of Chemistry, University of Patras, Caratheodory 1, University Campus, 26504 Patras, Greece; k.andrikopoulos@ac.upatras.gr (K.C.A.); andreopo@upatras.gr (A.K.A.); kallitsi@upatras.gr (J.K.K.)
[3] Organic Electronic Technologies P.C. (OET), 20th KM Thessaloniki—Tagarades, 57001 Thermi, Greece
* Correspondence: detselek@physics.auth.gr (D.T.); mgiot@physics.auth.gr (M.G.); Tel.: +30-2310-998103 (M.G.)

Citation: Tselekidou, D.; Papadopoulos, K.; Andrikopoulos, K.C.; Andreopoulou, A.K.; Kallitsis, J.K.; Logothetidis, S.; Laskarakis, A.; Gioti, M. Optical, Photophysical, and Electroemission Characterization of Blue Emissive Polymers as Active Layer for OLEDs. *Nanomaterials* **2024**, *14*, 1623. https://doi.org/10.3390/nano14201623

Academic Editor: Zhixing Gan

Received: 7 September 2024
Revised: 6 October 2024
Accepted: 8 October 2024
Published: 10 October 2024

Copyright: © 2024 by the authors. Licensee MDPI, Basel, Switzerland. This article is an open access article distributed under the terms and conditions of the Creative Commons Attribution (CC BY) license (https:// creativecommons.org/licenses/by/ 4.0/).

Abstract: Polymers containing π-conjugated segments are a diverse group of large molecules with semiconducting and emissive properties, with strong potential for use as active layers in Organic Light-Emitting Diodes (OLEDs). Stable blue-emitting materials, which are utilized as emissive layers in solution-processed OLED devices, are essential for their commercialization. Achieving balanced charge injection is challenging due to the wide bandgap between the HOMO and LUMO energy levels. This study examines the optical and photophysical characteristics of blue-emitting polymers to contribute to the understanding of the fundamental mechanisms of color purity and its stability during the operation of OLED devices. The investigated materials are a novel synthesized lab scale polymer, namely poly[(2,7-di(p-acetoxystyryl)-9-(2-ethylhexyl)-9H-carbazole-4,4′-diphenylsulfone)-co-poly(2,6-diphenylpyridine-4,4′-diphenylsulfone] (CzCop), as well as three commercially supplied materials, namely Poly(9,9-di-n-octylfluorenyl-2,7-diyl) (PFO), poly[9,9-bis(2′-ethylhexyl) fluorene-2,7-diyl] (PBEHF), and poly (9,9-n-dihexyl-2,7-fluorene-alt-9-phenyl-3,6-carbazole) (F6PC). The materials were compared to evaluate their properties using Spectroscopic Ellipsometry, Photoluminescence, and Atomic Force Microscopy (AFM). Additionally, the electrical characteristics of the OLED devices were investigated, as well as the stability of the electroluminescence emission spectrum during the device's operation. Finally, the determined optical properties, combined with their photo- and electro-emission characteristics, provided significant insights into the color stability and selectivity of each material.

Keywords: blue-light emitting OLEDs; polyfluorene; polycarbazole; spectroscopic ellipsometry; color selectivity

1. Introduction

To date, electrically semiconducting polymers have proved to be irreplaceable materials in the development of various electronic and optical devices, such as Organic Light-Emitting Diodes (OLEDs), Organic Field Effect Transistors (OFETs), Organic Photovoltaics (OPVs) etc. Among the main advantages are their solubility in common solvents, mechanical flexibility, non-expensive fabrication, and processing, with conductivity levels comparable to those of inorganic semiconductors or even metals [1–6]. Therefore, understanding their basic properties is essential for the design of novel semiconducting polymers, which are used as emissive layers and applied in solution-processed OLED devices. Wet-based deposition techniques are the most attractive methods for achieving flexible large-area full-color displays at a low cost due to their compatibility with roll-to-roll fabrication [7–10]. In particular, the emergence of flexible and wearable electronic devices as a part of the Internet of Things could be an important driving force to a new commercialization area of these materials, semiconducting polymers capable of responding adequately to these devices' requirements [11].

Notably, there is a significant need for research on improved conjugated polymers towards the three primary colors. It is well known that OLED devices require red, green, and blue emissions with high stability, efficiency, and color purity. It is important to mention that red and green light-emitting materials have been exhibiting excellent luminous efficiency and spectral stability [3,12–16]. On the other hand, developing blue light-emitting materials presents a significant challenge, leading to increased research interest in producing stable and high-quality blue light [17,18]. The intrinsic wide band gaps of blue emissive materials result in a high charge injection barrier and unbalanced injection and transportation of charges [15].

Specifically, the conjugated polymers containing the Poly-fluorene (PF) and Polycarbazole (PCz) motifs are promising candidates for such blue-light emitting materials. These materials have garnered increased attention owing to their good electro- and photoactive properties, as well as their high hole transporting mobility and strong absorption in the UV spectral region [11]. These materials also provide good thin film morphology and, for this reason, are promising candidates for application as an emissive layer in OLED devices [8,14,16]. In addition, they can be utilized as host materials for internal color conversion in blends with other conjugated polymers and with phosphorescent dyes providing easy color tuning [14,19–22].

However, fluorene-based derivatives may suffer from poor color purity and stability. Under prolonged device operation or annealing of the materials in air, PF-type materials often appear to degrade, resulting in long-range emission at photon ranges of 2.2–2.3 eV. There are two possible mechanisms that have been proposed and intensively debated to explain the origin of the undesirable emission band centered at 540–550 nm: (i) excimer emission due to interchain aggregation, or (ii) fluorenone formation in 9-position due to oxidation. For the first possible mechanism, initially, reordering of the polymer chains and subsequent aggregation, as well as excimer formation, was assigned as the source of the green emission. Instead of that, the second possible phenomenon was associated with on-chain defects incorporated during synthesis. Their oxidation leads to the presence of ketone defects, yielding the so-called fluorenone moieties incorporated into the polymer backbone. Currently, it is widely agreed that ketone defects are responsible for green emissions. As a result, the color of the emission shifts from the desired blue to the blue-green region (or even yellow). In order to improve the performance of PF-based OLED devices, it is important to identify the origin of the red-shift emission and to understand the mechanism of color degradation [6,23–29].

To realize improved performance, it is essential to synthesize new blue-light emitting materials with high color stability and selectivity for blue OLEDs. Significant efforts have been made to design versatile blue fluorescent materials aiming at further improving device efficiency, chromaticity, and lifetime. To this end, the combination of the synthesis of novel blue-emitting polymers and simple wet-fabricated OLEDs remains an open issue in the research field of OLEDs.

In this work, we present the comparative study of the optical and photophysical properties of a new synthesized lab-scale polyethersulfone, namely poly[(2,7-di (p-acetoxystyryl)-9-(2-ethylhexyl)-9H-carbazole-4,4'-diphenylsulfone)-co-poly(2,6-diphenylpyrydine- 4,4'-diphenylsulfone] (CzCop), with three commercially supplied blue-emitting polymers, the poly(9,9-di-n-octylfluorenyl-2,7-diyl) (PFO), the poly[9,9-bis(2'-ethylhexyl) fluorene-2,7-diyl] (PBEHF), and the poly (9,9 n-dihexyl-2,7-fluorene-alt-9-phenyl-3,6 carbazole) (F6PC). These emissive polymers consist mainly of fluorene and carbazole units. Their derivatives are based on alternating the fluorene and carbazole units or modifying the main chain with side groups in order to achieve better solubility in common solvents and film-forming ability. The commercially available fully conjugated blue-emitting polymers are directly compared with CzCop, which, apart from the carbazole moiety, also differentiates itself in that it incorporates a polyether motif in the main chain, disrupting the conjugation of the emitting moieties with oxygen heteroatoms. Specifically, aromatic polyethersulfones have been investigated as potential polymeric materials to be used in the emissive layer of OLEDs, incorporating fluorescent or even phosphorescent moieties, displaying attractive properties in terms of their facile synthesis, easier purification, film-forming ability, and ease of processability [30,31]. These materials could be applied in solution-processed OLED devices as emissive layers. The fabricated OLEDs have been subsequently studied and characterized in terms of their electroluminescence properties. This work aims to fully define and compare the innovative lab scale polymer with the commercially available ones, focusing on the selective stable blue emission. Determining the materials' optical properties in combination with the devices' photo- and electro-emission characteristics provides us with the necessary information to investigate and discuss the possibilities for applying these polymers to the proposed blue OLED devices, and the first encouraging results are obtained.

2. Materials and Methods

2.1. Materials

Copolymer CzCop were synthesized according to previously published procedures described elsewhere [27]. Poly(9,9-di-n-octylfluorenyl-2,7-diyl), PFO (Mw = 114,050) was supplied by Ossila (Sheffield, UK), whereas poly[9,9-bis(2′-ethylhexyl) fluorene-2,7-diyl], PBEHF (Mw = 79,000) and poly (9,9 n-dihexyl-2,7-fluorene-alt-9-phenyl-3,6 carbazole) F6PC (Mw = 9195) were supplied by Sigma Aldrich Chemie GmbH (Taufkirchen, Germany).

2.2. Ink Formulation

For the Hole Transport Layer (HTL), a solution of poly-3,4-ethylene dioxythiophene: poly-styrene sulfonate (PEDOT:PSS, Clevios Heraus Germany, Leverkusen, Germany) AI 4083 mixed with ethanol in the ratio of 2:1 was prepared. The PFO, PBEHF, and F6PC polymers were dissolved in chloroform with a resulting concentration of 1% wt. The synthesized copolymer CzCop (Mw = 69,000) was dissolved in N,N-Dimethylformamide (DMF) with a consequent concentration of 1% wt. The chemical structures for all studied polymers are depicted in Scheme 1.

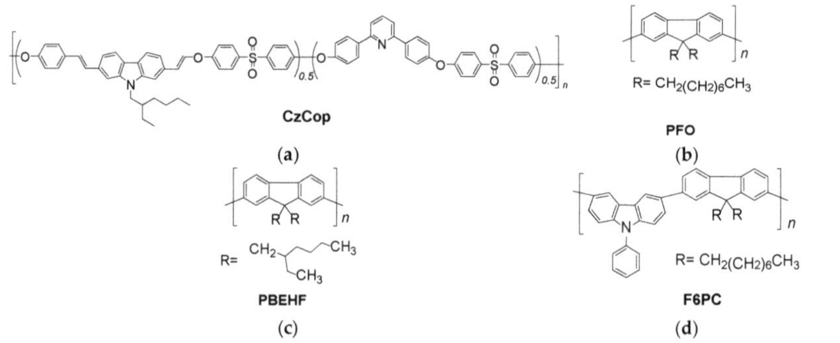

Scheme 1. Chemical structure of emitting polymer: (a) CzCop, (b) PFO, (c) PBEHF, and (d) F6PC.

2.3. OLED Fabrication

The fabricated OLED devices are structured as shown in Scheme 2. Firstly, pre-patterned Indium-Tin Oxide-coated (ITO) glass substrates (received by Ossila, Sheffield, UK) were extensively cleaned by sonication in DI, acetone, and ethanol for 10 min, followed by drying under nitrogen. The substrates were also treated with oxygen plasma at 40 W for 3 min. Then, the PEDOT:PSS layer, which was used as the HTL, was deposited by the spin coating method onto the glass/ITO substrate, followed by annealing at 120 °C for 5 min. The emitting layers (EML) were spun at the same speed, 2000 rpm/s for 60 s, onto the

PEDOT:PSS layer. Finally, a bilayer of Ca with a thickness of 6 nm and Ag with a thickness of 125 nm, which was used as an electron transport layer and cathode, respectively, was deposited using the appropriate shadow masks by Vacuum Thermal Evaporation (VTE).

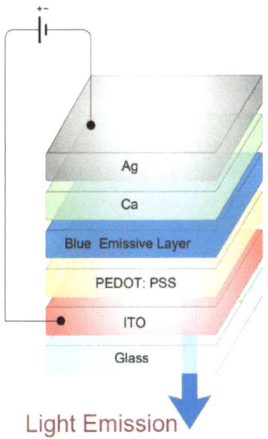

Scheme 2. The architecture of the fabricated devices.

2.4. Thin Film and Device Characterization

Spectroscopic Ellipsometry (SE) is a powerful and robust, non-destructive, and surface-sensitive optical technique for the determination of the optical properties as well as the thickness of the light-emitting polymers. Through the SE technique, we can measure the pseudodielectric function of the studied thin films. Moreover, by employing suitable modeling and fitting procedures, we can obtain valuable information regarding the dielectric function $\varepsilon(\omega)$, the precise thickness of the nanometer-scale thin films, the absorption coefficient, and optical constants such as the fundamental band gap and absorption energies (optical gaps). The SE measurements were conducted using a phase-modulated ellipsometer (UVISEL JobinYvon, Horiba Europe Research Center, Palaiseau, France) from the near IR to far UV spectral region 1.5–6.5 eV with a step of 20 meV at 70° angle of incidence. The SE experimental data were fitted to model-generated data using the Levenberg–Marquardt algorithm, which took into consideration all the fitting parameters of the applied model.

$$\langle \varepsilon(\omega) \rangle = \langle \varepsilon_1(\omega) \rangle + i \langle \varepsilon_2(\omega) \rangle \tag{1}$$

The surface morphology of the emitting thin films was investigated by Atomic Force Microscopy (AFM) (NTEGRA, NT-MDT, Moscow, Russia) in ambient conditions, using the tapping scanning mode and silicon-based cantilevers with a high-accuracy conical tip and nominal tip roundness < 10 nm.

Finally, the Photoluminescence (PL) and Electroluminescence (EL) characteristics of the active layers and of the final OLED devices, respectively, were measured using the Hamamatsu Absolute PL Quantum Yield measurement system (C9920-02, Jokocho, Higashi-ku, Hamamatsu City, 431-3196, Japan) and the external quantum efficiency system (C9920-12, Jokocho, Higashi-ku, Hamamatsu City, 431-3196, Japan), which measures brightness and light distribution of the devices. The current density-voltage and the luminance-voltage characteristics of the devices were measured using the Electroluminescence technique.

3. Results

3.1. Spectroscopic Ellipsometry

The optical and electronic properties of the blue-light emitting polymers were determined by modeling and analyzing the measured complex pseudo-dielectric function,

$\langle\varepsilon(E)\rangle$, via SE in the visible to far ultraviolet (Vis-fUV) spectral region. To obtain quantitative information from the measured $\langle\varepsilon(E)\rangle$ spectra, this has been analyzed by the use of a 5-phase theoretical model consisting of the layer sequence air/blue-light emitting polymer/PEDOT:PSS/ITO/Glass. We sequentially measured each layer and used the appropriate theoretical optical model to calculate the optical constants and thickness of each layer.

Figure 1a,b show the experimentally measured real part $\langle\varepsilon_1(E)\rangle$ and the imaginary part $\langle\varepsilon_2(E)\rangle$ (symbols) spectra of the pseudo-dielectric function $\langle\varepsilon(E)\rangle$, as a function of the photon energy in the range of 1.5–6.5 eV, as well as the corresponding fitted ones (dash lines). Specifically, for the determination of the dielectric response of the emitting polymers PFO, PBEHF, F6PC, and CzCop, we have used the modified Tauc–Lorentz (TL) dispersion oscillator model, which has been successfully applied in amorphous organic semiconductors [32–35]. The TL model is a powerful tool that can accurately describe interband absorptions above the energy bandgap and is presented in detail in previous works [32,36]. The imaginary part of the dielectric function, which is directly related to electronic absorption, is given by the expressions:

$$\varepsilon_2(E) = \sum_i \frac{A_i E_{0i} \Gamma_i \left(E - E_g^{TL}\right)^2}{\left(E^2 - E_{0i}^2\right)^2 + \Gamma_i^2 E^2} \cdot \frac{1}{E}, \; E > E_g^{TL} \tag{2}$$

$$\varepsilon_2(E) = 0, \; E \leq E_g^{TL} \tag{3}$$

where E_g^{TL} is the energy band gap, E_{0i} the resonance energy, Γ_i the broadening, and A_i the strength of the ith oscillator.

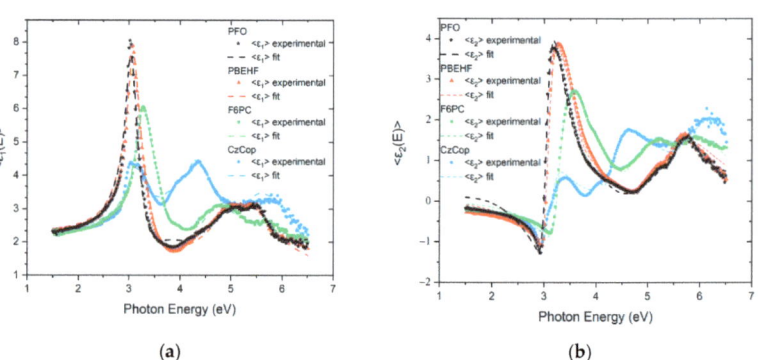

Figure 1. The experimental (symbols) (a) real $\langle\varepsilon_1(E)\rangle$ and (b) imaginary $\langle\varepsilon_2(E)\rangle$ spectra of the pseudodielectric function and the corresponding fitted ones (lines) of the studied films, grown on Glass/ITO/PEDOT:PSS.

Figure 2a,b show the calculated real $\langle\varepsilon_1(E)\rangle$ and imaginary $\langle\varepsilon_2(E)\rangle$ parts of the dielectric function of the studied polymers derived from the best-fit analysis. Specifically, we used three (i = 3) TL oscillators for the commercially available polymers PFO and PBEHF, four (i = 4) and five (i = 5) TL oscillators for the F6PC and the synthesized CzCop, respectively, to accurately parameterize all the electronic transitions. The $\varepsilon(E)$ is determined based on the best-fit parameters from the analysis of the measured $\langle\varepsilon(E)\rangle$. This analysis provides insights into the thickness of the thin film, the energy of characteristic electronic transitions, and the energy band gap of the materials under investigation. The respective calculated values of these parameters are presented in Table 1.

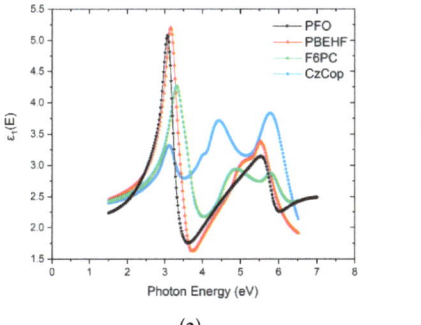
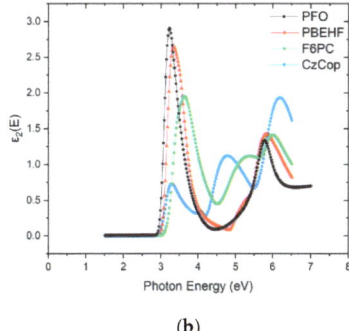

Figure 2. The (**a**) real $\varepsilon_1(E)$ and (**b**) imaginary $\varepsilon_2(E)$ part of the dielectric function $\varepsilon(E)$ of the studied films were calculated using the best-fit parameters derived by the SE analysis procedure.

Table 1. The calculated best-fit results were derived by the analysis of the measured $<\tilde{\varepsilon}(E)>$ spectra of the films.

	Thickness (nm)	Optical Band Gap (eV)		Electronic Transition Energy (eV)				
		E_g^{TL}	E_g^{Tauc}	E_{01}	E_{02}	E_{03}	E_{04}	E_{05}
PFO	47.0 ± 1	2.87	3.06	3.12			5.54	5.78
PBEHF	42.0 ± 1	2.95	3.11	3.16			4.73	5.66
F6PC	46.0 ± 1	2.99	3.28	3.36	3.50	4.47	5.69	
CzCop	27.0 ± 1	2.67	2.97	3.08	4.02	4.53	5.29	6.18

The dielectric response of blue-emitting materials is examined in relation to their electronic transition energies. For all studied polymers, the first electronic transition energy is calculated to be approximately $E_{01} \approx 3$ eV. However, the differences are evident and more pronounced at the higher energies range, above 3.5 eV. The PFO and PBEHF exhibit similarities in E_{04} and E_{05} energy values, which are significantly weaker in comparison to the E_{01}. The F6PC shows similar strength, but the E_{02} and E_{03} are calculated in lower energies. Finally, in the case of CzCop, it is a reduction in the strength of E_{01} compared to the following higher energies.

In the case of nanostructured amorphous polymeric films, the band gap E_g^{TL}, which is calculated using the TL dispersion equation and includes the apparent absorption resulting from disordering and localized defect states [32–36]. To accurately determine the fundamental band gap, which determines the color emission and is associated with the energy difference between the highest occupied molecular orbital (HOMO) and the lowest unoccupied molecular orbital (LUMO), we utilized the Tauc-plot method [35]. This method involves extrapolating the calculating $[E \times \varepsilon_2]^{1/2}$ to the zero ordinate to obtain a relatively wide band gap. The results of the E_g^{Tauc} are presented in Table 1 for comparison.

One can see that the PFO and PBEHF exhibit similar values for the first electronic absorption, which could be ascribed to the fluorene unit, whereas the F6PC presents different values of electronic absorption despite containing the fluorene unit in the polymer chain. In addition, the calculated values of the optical band gap E_g^{Tauc} for commercial polymers, PFO, PBEHF, and F6PC also present differences. Among the commercial polymers consisting of the main unit fluorene, the F6PC exhibits the higher value of the E_g^{Tauc}, which is equal to 3.28 eV. We can distinguish that the presence of the carbazole moiety affects the value of the band gap E_g^{Tauc} of F6PC, as it presents a higher value. This can be explained by the shortening of the conjugation length as the carbazole unit inserts into the conjugated backbone of PF, leading to an increase in the energy gap [9,37]. On the other hand, the two values of the band gap of CzCop present lower energies compared to the other ones and more electronic absorptions are observed with significantly broader characteristics.

Specifically, the calculated E_g^{Tauc} has the lowest value compared to the other three commercially supplied polymers and is calculated equal to 2.67 eV. This may be related to disorder-induced nanostructure and the formation of localized defect states in this film [33].

3.2. Absorption Coefficient and Photoluminescence

The Absorption Coefficient and PL spectra of thin films of PFO, PBEHF, F6PC, and CzCop are presented in Figure 3a, b, c, and d, respectively. The absorption coefficient spectrum of each emitting polymer is derived using the calculated bulk $\varepsilon(\omega)$. The spectra of the studied polymers show absorptions from 420 to 200 nm. In the case of the absorption coefficient spectrum of PFO, the dominant peak is located at 370 nm, which is attributed to the π-π* transition of the conjugated PF backbone [6,23]. The absorption coefficient spectrum of PBEHF shows a dominant maximum peak at around 360 nm, which is ascribed to π-π* transition [26,29]. The other PFO derivative containing the carbazole moiety, namely F6PC, exhibits a slightly blue-shifted absorption peak at 335 nm in comparison to PBEHF and PFO. This blue shift of the absorption peak can be attributed to a π-π* transition related to the presence of the carbazole unit. Sergent et al. presented the study of photophysical properties of blue-emitting fluorene-co-carbazole-based polymers, and they observed the blue shift of the absorption when the carbazole unit was incorporated along with fluorene [8]. They also proposed that this phenomenon is due to the interruption of the conjugation by the presence of 3,6 carbazole units within the conjugated main chain [8,37,38]. In the case of the novel copolymer CzCop, it is clear that the absorption spectrum shows a broad structureless band from 425 to 300 nm, centered at 380 nm, indicating that it exists in the disordered (amorphous) phase.

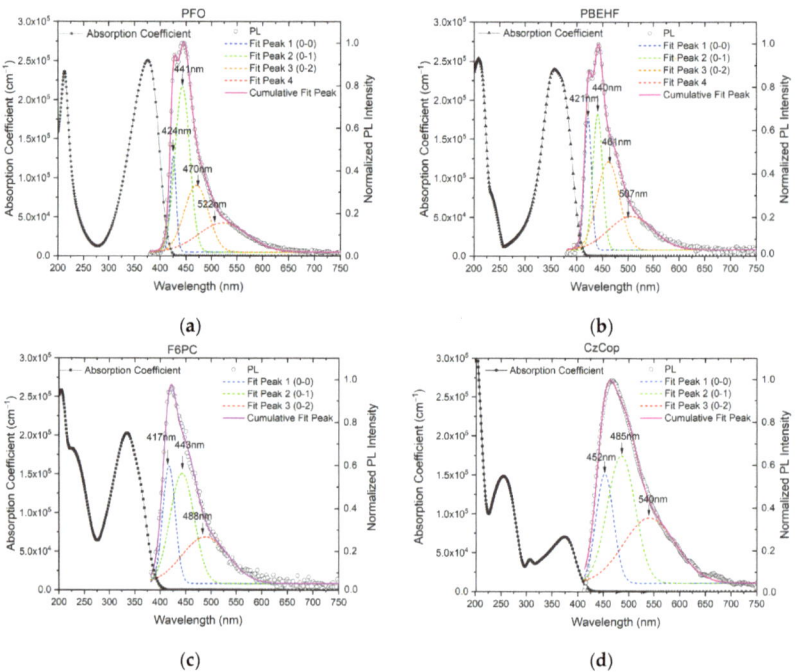

Figure 3. The absorption coefficient and PL emission spectra of (**a**) PFO, (**b**) PBEHF, (**c**) F6PC, and (**d**) CzCop.

Moreover, the emitting thin films were also examined by PL in order to define the luminescence behavior of the active materials. The PL spectra were recorded upon excitation at 370 nm. The right axes of Figure 3a, b, c, and d depict the PL spectra of PFO, PBEHF, F6PC, and CzCop, respectively. For a better evaluation of the PL emission spectra of the emitting films, a deconvolution fitting analysis of the experimental spectra was realized using a Gauss oscillator for the analysis procedure. The deconvolution analysis revealed that the PFO PL spectrum exhibits a vibronic structure with peaks at ~424 nm (0–0), ~441 nm (0–1), and ~470 nm (0–2); a fourth phonon side band (0–3) can also be seen at ~522 nm [6,39]. The PL spectrum of PBEHF is dominated by four distinct peaks, and notably, the PBEHF spectrum is typical of polyfluorenes, similar to the PL spectra of PFO. In particular, it shows a structured band with two sharp peaks at ~421 and ~440 nm, one at ~461 nm, and a minor shoulder at ~507 nm. Therefore, as mentioned above, the first three peaks correspond to an average vibronic progression of the 0–0, 0–1, and 0–2 intrachain singlet transitions, respectively [8,26,29,40,41]. On the other hand, it is remarkable to observe that the shape of the F6PC PL spectrum is different compared to the PFO and PBEHF. For the F6PC, the featureless PL emission was deconvoluted by three peaks located at ~417, ~443, and ~488 nm. The deconvolution analysis reveals a slight blue shift, especially from 421 or 424 to 417 nm. This fact makes us speculate that the carbazole unit affects the emission spectra of F6PC, as the emission is mostly governed by a radiative decay from the electronic states of the conjugated polymer comprising the carbazole unit [8,37,38]. This is due to the interruption of the delocalization of the π-electrons along the polymer backbone by the 3,6-carbazole linkages. Thus, as mentioned above, the presence of the carbazole unit indicates that the conjugation length is reduced, leading to an increased energy gap and resulting in a blue shift in PL emission. In addition, the PL emission of the CzCop presents a similar band shape compared to the F6PC PL spectra, as both F6PC and CzCop contain a carbazole unit in the main backbone. The spectrum of CzCop is shifted to longer wavelengths compared to the F6PC spectrum, as it is also indicated by the deconvolution analysis. This fact is assigned to conjugation disruption because of the ether linkages and structural disorders. The polyether backbone induces non-planar conformations that lead to more electronic transitions. Also, the electron-withdrawing/donating nature of the sulfone group and the Cz and Py groups could lead to more Charge Transfer states.

At the same time, the PL measurements are used to calculate the coordinates in the Commission Internationale de L' Eclairage (CIE) chromaticity diagram, as they are demonstrated in Figure 4. One can observe that the chromaticity coordinates are generally located in the spectral region of the blue region. More specifically, commercially available photoactive materials have coordinates situated in a region characterized by blue emission color [17]. For the PFO, PBEHF, and F6PC, the values of the corresponding coordinates are (0.17, 0.13), (0.17, 0.12), and (0.17, 0.13), respectively. On the other hand, concerning the synthesized polymer CzCop, its CIE Coordinates deviate from the deep blue region, and the emission can be characterized as sky-blue [17]. More specifically, the values of the CzCop chromatic coordinates are (0.19, 0.25).

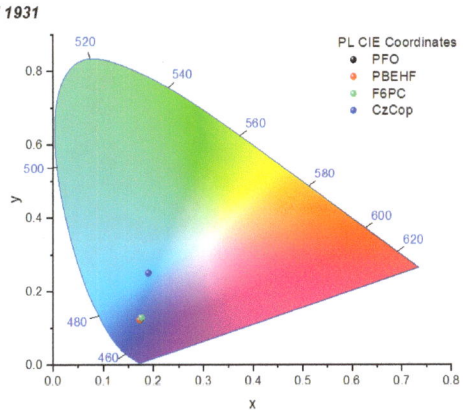

Figure 4. CIE diagram of PL emission of studied emitting materials.

3.3. Atomic Force Microscopy

Understanding interfaces is crucial for the further development and optimization of OLED devices as they encompass several thin film layers. In particular, the film morphology of the emissive layers is an important parameter that could be taken into account when introducing interfacial layers via wet techniques in optoelectronic devices. For this reason, we extensively studied the surface topography of the emissive thin films using AFM. Figure 5a, b, c, and d illustrate the AFM-measured surface topography (height) images of spin-coated thin films PFO, PBEHF, F6PC, and CzCop, respectively. The results derived from the AFM image analysis are presented in Table 2. It can be observed that the surface morphology of every sample was uniform and sufficiently covered the substrate. The image analysis revealed that almost smooth and continuous films were formed with low Root Mean Square (RMS) roughness values. More specifically, from the height distribution plots shown in Figure 6, it can be seen that most of the features detected are between 2 and 7 nm in height. However, it is noteworthy that the synthesized emitting polymer CzCop thin film exhibits the lowest RMS value compared to the other commercial polymers. The importance of this topic is related to the fact that the interfacial area between the thin films sets the condition for the injection of charges in a device, which has a profound influence on the device's operational characteristics, for example, the current-voltage characteristic. Generally, it has been established that smoother surfaces reduce the loss of the injection and transportation of charges at the interface, which is beneficial in OLED devices consisting of different layers [13,42]. Thus, the roughness of the emissive layer plays a significant role in the performance of the device and improves the charge injection in the optoelectronic devices, resulting in reduced turn-on voltages.

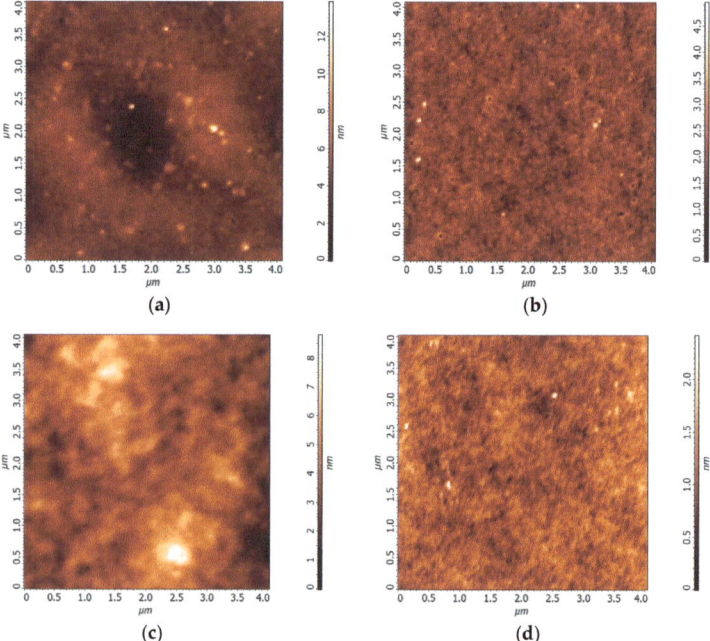

Figure 5. AFM topography Images Scan size 4 × 4 µm of (**a**) PFO, (**b**) PBEHF, (**c**) F6PC, and (**d**) CzCop.

Table 2. AFM results for the spin-coated thin films.

	Root Mean Square, Sq (nm)	Average Roughness, Sa (nm)	Peak to Peak, Sy (nm)
PFO	1.01	0.74	13.72
F6PC	0.86	0.67	8.67
PBEHF	0.32	0.25	4.94
CzCop	0.25	0.20	1.54

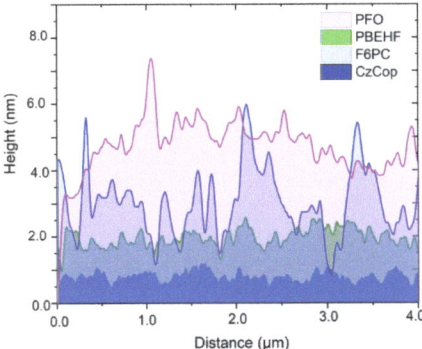

Figure 6. Cross section of AFM images of examined emissive thin films.

3.4. Electroluminescence

Electroluminescence preliminary investigations on the studied emissive materials were also carried out in order to evaluate their potentiality for the OLED technology. The EL spectra were recorded in the wavelength range from 380 to 900 nm by applying an external bias voltage from 3 to 14 V with 1 V step. In the case of CzCop, the maximum applied voltage was limited to 9 V due to the thinner active layer. Figure 7 shows the respective experimental EL spectra of the studied devices, which were recorded at 14 V for the commercial materials and at 8 V for the lab-scale synthesized polymer. The corresponding theoretical curves were obtained after the fitting deconvolution procedure using 4 Gauss oscillators for the PFO and F6PC, 5 Gauss oscillators for the PBEHF, and 3 Gauss oscillators for the CzCop. Table 3 summarizes the PL and EL emission spectra analysis, including the wavelengths of the λ^{max} peaks and the broadening FWHM.

Table 3. Results of the deconvolution analysis of the PL and EL emission spectra.

	Spectrum	PEAK 1		PEAK 2		PEAK 3		PEAK 4		PEAK 5	
		λ^{max} (nm)	FWHM (nm)	λ^{max} (nm)	FWHM (nm)	λ^{max} (nm)	FWHM (nm)	λ^{max} (nm)	FWHM (nm)	λ^{max} (nm)	FWHM (nm)
PFO	PL	424	12	441	28	470	47	522	96	-	-
	EL	429	13	454	18	484	20	503	88	-	-
PBEHF	PL	421	13	440	19	461	46	507	95	-	-
	EL	418	14	440	23	472	21	492	67	542	107
F6PC	PL	417	24	443	47	488	95	-	-	-	-
	EL	412	32	447	53	488	76	619	20	-	-
CzCop	PL	453	35	488	58	551	113	-	-	-	-
	EL	434	17	458	39	505	88	-	-	-	-

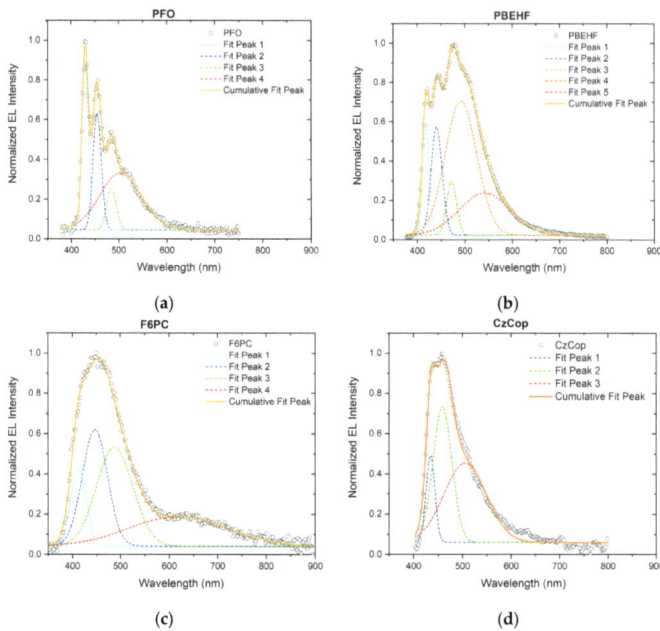

Figure 7. Deconvoluted normalized EL curves of (**a**) PFO, (**b**) PBEHF, (**c**) F6PC, and (**d**) CzCop.

The comparative study between the Electroluminescence and Photoluminescence spectra is shown in Figure 8 to better comprehend the emission characteristics. For the commercial light-emitting polymers PFO, PBEHF, and F6PC, it is obvious that the emission bands obtained from these studied devices are slightly broadened compared to their corresponding PL spectra. One can see that the maximum EL emission peaks of these commercial polymers are also moved to the longer wavelength region compared to their PL spectra. Specifically, from the EL spectrum of PFO can be observed a low-energy emission peak, which is located at approximately ~500 nm. The EL emission of PBEHF exhibits the dominant peak located at ~500 nm. As previously mentioned, the origin of the green-color emission in PF-type materials has been extensively debated in recent years. There are two possible explanations for this phenomenon. The first one is that the formation of a low-energy emission band at 2.2–2.3 eV occurred due to the formation of fluorenone defect sites (keto-defect) during the device operation, specifically in atmospheric conditions. In fact, it was shown that keto-defects form easily with the oxidation of monoalkylated fluorene monomer units during the device operation [23–26,43]. List et al. [24] presented in their study that the keto defects act as low-energy trapping sites for singlet excitons, being populated by an excitation energy transfer from the PF main chain. In particular, they found a much stronger contribution from the defect-related emission in EL than in PL spectra, which was attributed to two parallel processes: trapping of charges at the keto site and their subsequent emissive recombination in addition to energy transfer of singlet excitons from the PF main chain to keto defect sites. The other possible explanation is based on the aggregations and/or excimer formation in these materials, originating from the interchain attraction in the π-conjugated systems [44]. Since such interactions are short-range forces, the distance between the polymer chains is one of the governing factors for this phenomenon. As they become smaller, the polymer chains have a higher chance of entanglement with each other to form aggregates.

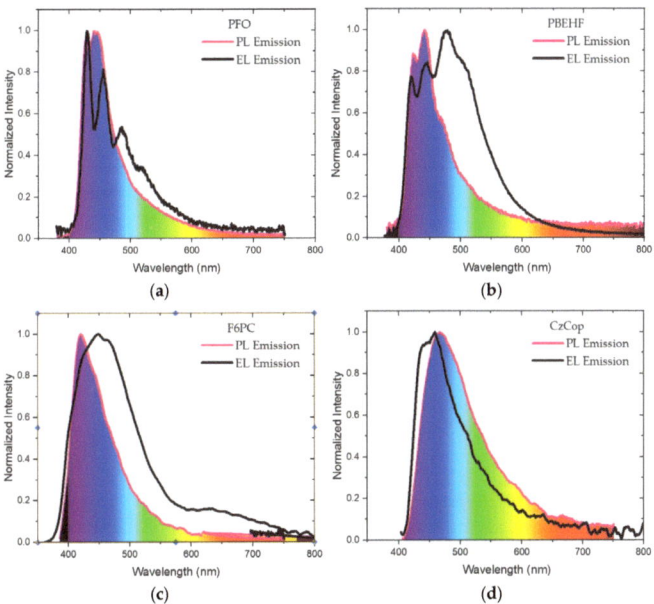

Figure 8. Comparison between EL and PL emission spectra of (**a**) PFO, (**b**) PBEHF, (**c**) F6PC, and (**d**) CzCop.

In the case of F6PC EL spectra, we can observe a dominant peak approximately at 455 nm and a shoulder at 650 nm. The latter probably originated from an electric-field-induced electromer emission. Deksnys et al. [45] observed EL emission in the long-wavelength region due to the electromer emission. In their study, they presented the synthesis, spectroscopic, thermal and electrochemical characterization of the ambipolar fluorophore 3,6-di(4,4′-dimethoxydipheny laminyl)-9-(1-naphthyl) carbazole (DPNC) which demonstrated a voltage-dependent green–blue electrofluorescence. It is well known that the electromer emission takes place mainly from the direct irradiative recombination of holes and electrons residing at two neighbouring molecules or from two molecules within some appropriate close distance.

Finally, deconvolution analysis verified that the novel emitting polymer CzCop has similar emission spectra for both EL and PL. However, we can observe the blue shift in EL spectra compared to PL because PL and EL mechanisms are different. It is well known that the PL emission is associated with the direct photoexcitation of the emissive thin film and the recombination from the excited states, whereas the EL depends on the carrier injection mechanisms, the transportation, and the recombination of charges across the device structure [32,36]. So, when comparing the PL and EL emission spectra of each emitting polymer, it is obvious that the novel lab-scale polymer CzCop exhibits the highest stability and color selectivity in its emission either as a photoactive thin film or as the active layer of an OLED device.

The CIE coordinates, derived from PL and EL measurements for the studied OLEDs, are illustrated in Figure 9. Generally, the CIE coordinates visualize the entire range of colors that can be obtained by mixing the three primary colors (red (R), green (G), and blue (B)) by varying the wavelength and emission intensity. According to the CIE diagram, it is obvious that the PL emission of commercial materials is approaching the blue region. On the other hand, in the case of lab-scale material, the PL CIE coordinates are located in the sky-blue region. Clearly, the PL CIE coordinates are different from the EL ones. Notably, the red shift of the EL emission spectra of the commercial materials is confirmed with the shift of CIE coordinates to the sky-blue region. As it is referred to above, the EL emission spectra exhibit different emission behavior compared to the PL. Specifically, the blue emission of the polymers turns into blue-green emission, as verified through the CIE diagram, except for the lab-scale material. The light of the CzCop device well approaches the blue region, making it a promising candidate as an efficient blue light-emitting material.

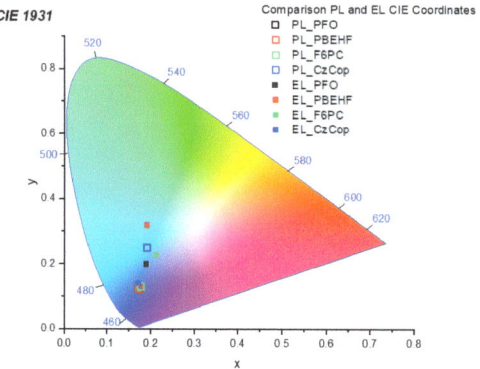

Figure 9. CIE diagram of PL and EL emission of the studied emitting films.

EL emission under different applied voltages was recorded to assess the crucial stability factor during OLED device operation. Figure 10a–d show the results obtained from the PFO, PBEHF, F6PC, and CzCop devices accordingly. Firstly, comparing the EL spectra of PFO for different applied voltages (1–14 V), an EL peak appeared at approximately ~500 nm, which increases with the applied voltage. Note that the additional featureless green emission may originate from the fluorenone moieties. This can be explained by the fact that the blue light originates from the bimolecular recombination of free electrons and holes, whereas the green light is generated by electrons that are trapped at fluorenone sites and subsequently recombine with a hole. So, the EL spectra of PFO are affected by the defects during the device operation.

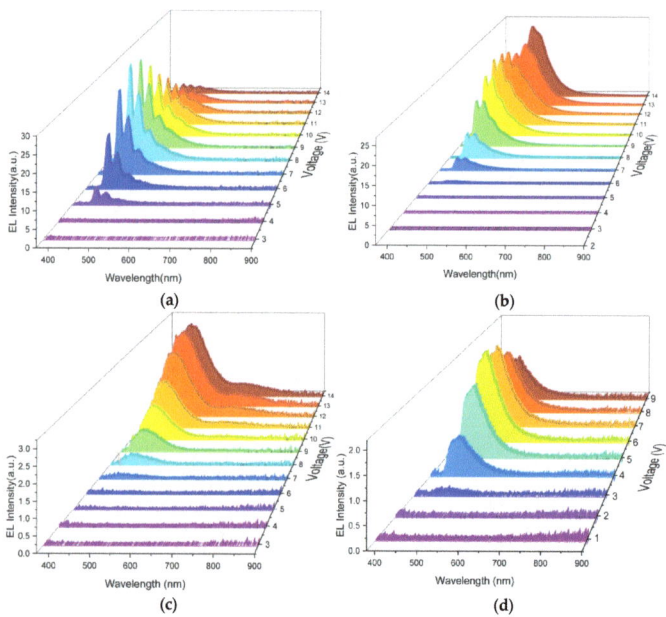

Figure 10. The evolution of EL emission spectra of the produced OLED devices, bearing as active film (**a**) PFO, (**b**) PBEHF, (**c**) F6PC, and (**d**) CzCop, under excitation with various bias voltages.

In addition, it can be seen that increases in the applied voltage (1–14 V) also induced changes to the EL spectral shape of PBEHF. At higher operation voltages, we can observe EL spectra with multiple peaks that span from the blue to the red region. The ratio of the intensity at 425 nm to 476 nm increased when the voltage increased from 7 to 12 V. When the applied voltage exceeded 12 V, another dominant peak at 507 nm emerged. As mentioned above, the presence of a peak at 507 nm is related to the defects of fluorene. More specifically, it is noteworthy that the green peak presents lower intensity relative to the blue peak, from 7 to 12 V. We can also assume that the blue light originates from the bimolecular recombination of free electrons and holes, whereas the green light is generated by charges that are trapped at the fluorenone defects and subsequently recombine. As a result, the blue emission will strongly increase with voltage, whereas the recombination from the trapping sites is limited by the number of traps [23]. However, when the applied voltage is increased, the intensity of a peak at 507 nm is increased as well. This can be explained by the fact that more charges are trapped at the fluorenone units due to the electrooxidation during the device operation.

For the case of the F6PC device (Figure 10c) and the respective EL spectra obtained under different applied voltages (1–14 V), it is obvious that when the applied voltage is increased from 7 to 10 V, the intensity of a dominant peak, centered at ~450 nm, is increased as well. When the voltage is increased above 10 V, the intensity of the shoulder emission peak located at 650 nm is also increased. This fact indicates that the electromer emission band demonstrates a clear dependence on the applied voltage. As the electromer-type excited states can only be formed under electrical excitation, their emission is visible only in the EL spectra.

Finally, the evolution of EL emission spectra of CzCop with various bias voltages (1–9 V) was also investigated (Figure 10d). It was derived that the EL spectra were nearly unchanged by increasing the driving voltages, indicating the high stability of CzCop in the EL process either as a luminophor or as the host material. This demonstrates that the newly synthesized material forms uniform coatings without defects that could create traps and, by extension, instability in the emission characteristics of the OLED devices to which they are applied. Thus, CzCop exhibits superior EL stability and emission in the blue region, according to the CIE diagram, compared to the other commercially available materials.

The EL spectra of PFO, PBEHF, F6PC, and CzCop OLEDs are converted to the CIE coordinate and overlaid onto the approximate color regions on a CIE 1931 (x, y) chromaticity diagram. The evolution of EL chromaticity coordinates as a function of the applied voltage is illustrated in Figure 11a–d for each polymer. According to the CIE diagram, the emissions from PFO and PBEHF OLEDs were at the edge of the blue-violet region at 8 V, but as the voltage increased, both were emitted to the greenish-blue region. This behavior can be explained as the resultant emission shifted when the intensity of the approximately ~490 nm peak increased. Quite interestingly, the EL chromaticity coordinates of PBEHF present a higher shift compared to the PFO, and this fact can be attributed to the peak emission at ~490 nm, which is the dominant peak at the EL emission, as shown in Figure 11b. For the device based on F6PC, the emissions in the blue area shift towards the center region with increasing voltage, largely due to the appearance of the red emission at 600 nm. Notably, as the applied voltage increases, the novel material CzCop exhibits superior emission stability, independent of the driving voltage. Therefore, the lab-scale polymer CzCop demonstrates excellent stability and color selectivity during the device operation, making it a promising candidate as an efficient blue light-emitting material.

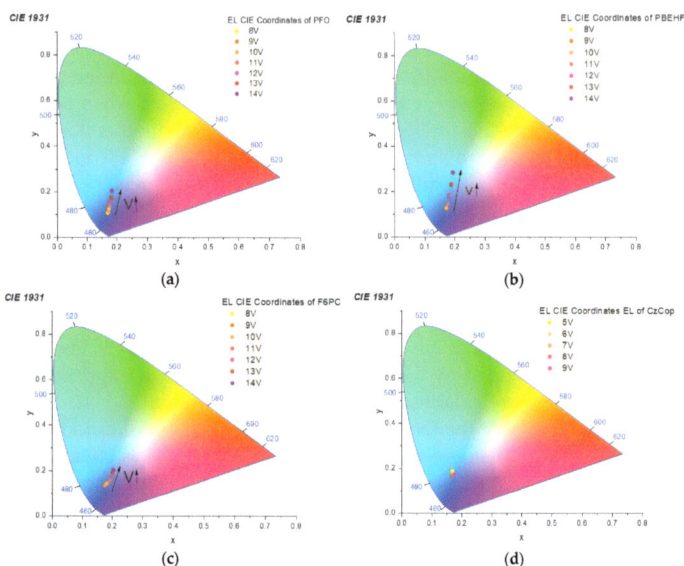

Figure 11. The evolution of EL Chromaticity Coordinates of the produced OLED devices, bearing as active film (**a**) PFO, (**b**) PBEHF, (**c**) F6PC, and (**d**) CzCop, under excitation with various bias voltages. Arrows show the shift of color coordinates towards the sky-blue region with increasing Voltage (V).

3.5. Electrical Characteristics

In terms of electrical characteristics, Figure 12a and b show the current–voltage and luminance–voltage characteristics of the fabricated devices on a logarithmic scale, respectively. The electrical characteristics and chromaticity coordinates, which were derived from the EL spectra, are summarized in Table 4. One can see that the novel polymer CzCop exhibits a lower turn-on voltage and lower potential operating voltage compared to the other commercially supplied polymers. It is also important to observe that the novel lab-scale material approaches the electrical characteristics of the commercially supplied ones. The measured luminance at 8 V for the PFO, PBEHF, F6PC, and CzCop is 413, 62, 3, and 28 cd/m^2, respectively. The maximum luminance measured for the commercial polymers at higher bias voltages exhibits higher values, but this can be ascribed to the fact that their thicknesses are higher compared to the lab-scale polymer CzCop.

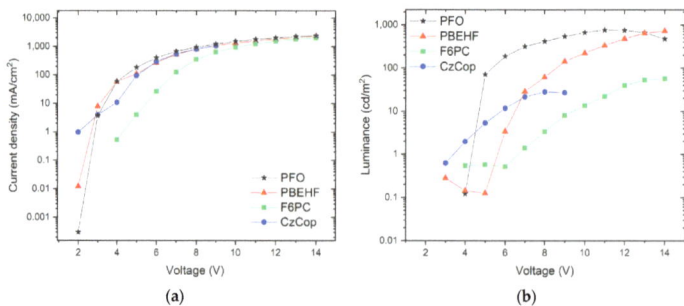

Figure 12. Logarithmic plot of (**a**) Current density–Voltage (J–V) and (**b**) Luminance–Voltage (L–V) characteristic curves of the fabricated OLED devices.

Table 4. The electrical–operational characteristics and EL CIE coordinates of fabricated OLED devices.

	Current Density Turn-On Voltage (V)	Luminance Turn-On Voltage (V)	Luminance at 8 V (cd/m^2)	Luminance (cd/m^2)	CIEx	CIEy
PFO	2.7	4.0	413	759	0.18	0.20
PBEHF	2.8	5.6	62	729	0.19	0.32
F6PC	4.2	6.6	3	57	0.21	0.23
CzCop	2.0	3.4	28	28	0.16	0.16

Nevertheless, a more extensive and in-depth optimization of the device architecture is essential to enhance overall efficiency. Since it is a novel material, additional device architecture optimization is required in terms of the precise selection and alignment of the Hole Transport Layer-Hole Injection Layer and Electron Injection Layer-Electron Transport Layer [46]. The identification of the most suitable functional materials for each layer in the OLED architecture, along with the smooth surface of the CzCop film that enhances interfacial properties, is expected to result in the production of more efficient devices. Overall, this comparison demonstrates that the commercial materials exhibited undesired blue-green emissions, while the CzCop synthetic polymer nicely supports color stability and selectivity in the pure blue region during device operation, thus paving the way for use in the emerging field of OLEDs. Thus, the CzCop offers promising potential for applications, including displays and medical devices. For the latter, "blue light therapy" has a range of beneficial effects [47].

4. Conclusions

In this study, we investigated the optical properties of four different blue light-emitting polymeric materials. Three of these materials (PFO, PBEHF, and F6PC) were commercially available, while the fourth one (CzCop) was a novel polymer synthesized in the laboratory. These materials have been applied as emissive layers in the wet fabrication of OLED devices using the spin coating process. In terms of the optical characterization, a thorough comparison between these materials concerning their thickness, dielectric function, fundamental energy gap, and absorption coefficient was obtained using Spectroscopic Ellipsometry. In addition, quantitative analysis of the PL and EL emission peaks and widths was performed for all the studied materials to evaluate their color stability and selectivity. It was realized that the lab-scale polymer CzCop exhibits superior color stability and selectivity during the device operation. On the other hand, the commercially available PFO, PBEHF, and F6PC show variations in their maximum peak emission wavelength and red shift in their EL spectra. Thus, CzCop is a promising candidate as a blue light-emitting material compared to other commercial polymers. It has great potential for achieving a stable blue color in various high-emergence applications. However, further investigation is required to enhance the functionalization of the fabricated WOLED devices in order to achieve greater efficiency.

Author Contributions: Conceptualization, writing—original draft preparation, visualization, formal analysis, data curation D.T.; methodology, investigation, data curation K.P.; investigation, data curation, validation, writing—review and editing K.C.A.; investigation, methodology, validation, writing—review and editing A.K.A.; conceptualization, funding acquisition, methodology J.K.K.; conceptualization, resources S.L.; investigation, data curation A.L.; supervision, conceptualization, visualization, writing—review and editing, funding acquisition M.G. All authors have read and agreed to the published version of the manuscript.

Funding: This research received no external funding.

Data Availability Statement: Data presented in this article is available on request from the corresponding author.

Acknowledgments: This work was supported by the European Regional Development Fund of the European Union and Greek national funds through the Operational Program Competitiveness,

Entrepreneurship, and Innovation under the call RESEARCH—CREATE—INNOVATE (project code: T1EDK-01039) and the COPE-Nano EU project (Grant Agreement No. 101059828).

Conflicts of Interest: Author Stergios Logothetidis was employed by the company Organic Electronic Technologies P.C. (OET). The remaining authors declare that the research was conducted in the absence of any commercial or financial relationships that could be construed as a potential conflict of interest. The funders had no role in the design of the study, in the collection, analyses, or interpretation of data, in the writing of the manuscript, or in the decision to publish the results.

References

1. Namsheer, K.; Rout, C.S. Conducting polymers: A comprehensive review on recent advances in synthesis, properties and applications. *RSC Adv.* **2021**, *11*, 5659–5697. [CrossRef]
2. Lim, J.W. Polymer Materials for Optoelectronics and Energy Applications. *Materials* **2024**, *17*, 3698. [CrossRef] [PubMed]
3. Andreopoulou, A.K.; Gioti, M.; Kallitsis, J.K. Organic Light-emitting Diodes Based on Solution-Processable Organic Materials. In *Solution-Processable Components for Organic Electronic Devices*, 1st ed.; Łuszczynska, B., Matyjaszewski, K., Ulanski, J., Eds.; Wiley-VCH Verlag GmbH & Co. KGaA: Weinheim, Germany, 2019; Volume 8, pp. 413–482.
4. Zhu, M.; Yang, C. Blue fluorescent emitters: Design tactics and applications in organic light-emitting diodes. *Chem. Soc. Rev.* **2013**, *42*, 4963–4976. [CrossRef] [PubMed]
5. Tankelevičiūtė, E.; Samuel, I.D.; Zysman-Colman, E. The Blue Problem: OLED Stability and Degradation Mechanisms. *J. Phys. Chem. Lett.* **2024**, *15*, 1034–1047. [CrossRef]
6. Zhang, Q.; Wang, P.-I.; Ong, G.L.; Tan, S.H.; Tan, Z.W.; Hii, Y.H.; Wong, Y.L.; Cheah, K.S.; Yap, S.L.; Ong, T.S. Photophysical and Electroluminescence Characteristics of Polyfluorene Derivatives with Triphenylamine. *Polymers* **2019**, *11*, 840. [CrossRef]
7. Colella, C.A.; Griffin, M.; Kingsley, J.; Scarratt, J.; Luszczynska, N.; Ulanski, J.B. Slot-Die Coating of Double Polymer Layers for the Fabrication of Organic Light Emitting Diodes. *Micromachines* **2019**, *10*, 53. [CrossRef]
8. Sergent, A.; Zucchi, G.; Pansu, R.B.; Chaigneau, M.; Geffroy, B.; Tondelier, D.; Ephritikhine, M. Synthesis, characterization, morphological behaviour, and photo- and electroluminescence of highly blue-emitting fluorene-carbazole copolymers with alkyl side-chains of different lengths. *J. Mater. Chem. C* **2013**, *1*, 3207–3216. [CrossRef]
9. Cook, J.H.; Santos, J.; Al-Attar, H.A.; Bryce, M.R.; Monkman, A.P. High brightness deep blue/violet, fluorescent polymer light-emitting diodes (PLEDs). *J. Mater. Chem. C* **2015**, *3*, 9664–9669. [CrossRef]
10. Bail, R.; Hong, J.Y.; Chin, B.D. Inkjet printing of blue phosphorescent light emitting layer based on bis(3,5-di(9H-carbazol-9-yl))diphenylsilane. *RSC Adv.* **2018**, *8*, 11191–11197. [CrossRef]
11. Bekkar, F.; Bettahar, F.; Moreno, I.; Meghabar, R.; Hamadouche, M.; Hernáez, E.; Vilas-Vilela, J.L.; Ruiz-Rubio, L. Polycarbazole and Its Derivatives: Synthesis and Applications. A Review of the Last 10 Years. *Polymers* **2020**, *12*, 2227. [CrossRef]
12. Fallahi, A.; Taromi, F.A.; Mohebbi, A.; Yuenc, J.D.; Shahinpoor, M. A novel ambipolar polymer: From organic thin-film transistors to enhanced air-stable blue light emitting diodes. *J. Mater. Chem. C* **2014**, *2*, 6491–6501. [CrossRef]
13. Sun, J.; Zhang, T.; Liao, X.; Wang, K.; Hou, M.; Wu, D.; Miao, Y.; Wang, H.; Bingshe, X. A novel luminophor and host polymer from fluorene-carbazole derivatives for preparing solution-processed non-doped blue and closed-white light devices. *Tetrahedron* **2018**, *74*, 1053–1058. [CrossRef]
14. Lv, X.; Sun, M.; Xu, L.; Wang, R.; Zhou, H.; Pan, Y.; Zhang, S.; Sun, O.; Xue, S.; Yang, W. Highly efficient non-doped blue fluorescent OLEDs with low efficiency roll-off based on hybridized local and charge transfer excited state emitters. *Chem. Sci.* **2020**, *11*, 5058–5506. [CrossRef]
15. Das, D.; Gopikrishna, P.; Singh, A.; Dey, A.; Iyer, P.K. Efficient blue and white polymer light emitting diodes based on a well charge balanced, core modified polyfluorene derivative. *Phys. Chem. Chem. Phys.* **2016**, *18*, 7389–7394. [CrossRef]
16. Liu, R.; Xiong, Y.; Zeng, W.; Wu, Z.; Du, B.; Yang, W.; Sun, M.; Cao, Y. Extremely Color-Stable Blue Light-Emitting Polymers Based on Alternating 2,7-Fluorene-co-3,9-carbazole Copolymer. *Macromol. Chem. Phys.* **2007**, *208*, 1503–1509. [CrossRef]
17. Siddiqui, I.; Kumar, S.; Tsai, Y.-F.; Gautam, P.; Shahnawaz; Kesavan, K.; Lin, J.-T.; Khai, L.; Chou, K.-H.; Choudhury, A.; et al. Status and Challenges of Blue OLEDs: A Review. *Nanomaterials* **2023**, *13*, 2521. [CrossRef]
18. Huang, T.; Wang, Q.; Zhang, H.; Xin, Y.; Zhang, Y.; Chen, X.; Zhang, D.; Duan, L. Delocalizing electron distribution in thermally activated delayed fluorophors for high-efficiency and long-lifetime blue electroluminescence. *Nat. Mater.* **2024**. [CrossRef] [PubMed]
19. Shi, Y.; Wang, Z.; Meng, T.; Yuan, T.; Ni, R.; Li, Y.; Li, X.; Zhang, Y.; Tan, Z.; Lei, S.; et al. Red Phosphorescent Carbon Quantum Dot Organic Framework-Based Electroluminescent Light-Emitting Diodes Exceeding 5% External Quantum Efficiency. *J. Am. Chem. Soc.* **2021**, *143*, 18941–18951. [CrossRef] [PubMed]
20. Zhang, Y.; Feng, N.; Zhou, S.; Xin, X. Fluorescent nanocomposites based on gold nanoclusters for metal ion detection and white light emitting diodes. *Nanoscale* **2021**, *13*, 4140–4150. [CrossRef]
21. Sun, P.; Wang, Z.; Sun, D.; Bai, H.; Zhu, Z.; Bi, Y.; Zhao, T.; Xin, X. pH-guided self-assembly of silver nanoclusters with aggregation-induced emission for rewritable fluorescent platform and white light emitting diode application. *J. Colloid Interface Sci.* **2020**, *567*, 235–242. [CrossRef]

22. Xie, H.; Wang, Y.; Zhang, N.; Li, S.; Li, J.; Xin, X. Solvent-Induced Self-Assembly of Silver Nanoclusters for White-Light-Emitting Diodes and Temperature Sensing. *ACS Appl. Nano Mater.* **2024**, *7*, 1009–1018. [CrossRef]
23. Kuik, M.; Wetzelaer, G.-J.A.H.; Laddé, J.G.; Nicolai, H.T.; Wildeman, J.; Sweelssen, J.; Blom, P.W.M. The Effect of Ketone Defects on the Charge Transport and Charge Recombination in Polyfluorenes. *Adv. Funct. Mater.* **2011**, *21*, 4502–4509. [CrossRef]
24. List, E.J.W.; Gaal, M.; Guentner, R.; Freitas, P.S.; Scherf, U. The role of keto defect sites for the emission properties of polyfluorene-type materials. *Synth. Met.* **2003**, *139*, 759–763. [CrossRef]
25. Chew, K.W.; Abdul Rahim, N.A.; Teh, P.L.; Osman, A.F. Enhanced luminescence stability and oxidation–reduction potential of polyfluorene for organic electronics: A review. *Polym. Bull.* **2024**, *81*, 7659–7685. [CrossRef]
26. Hwang, D.-H.; Park, M.-J.; Lee, J.-H.; Cho, N.-S.; Shim, H.-K.; Lee, C. Synthesis and light-emitting properties of polyfluorene copolymers containing a hydrazone derivative as a comonomer. *Synth. Met.* **2004**, *146*, 145–150. [CrossRef]
27. Chochos, C.L.; Kallitsis, J.K.; Georgiou, V.G. Rod-Coil Block Copolymers Incorporating Terfluorene Segments for Stable Blue Light Emission. *J. Phys. Chem. B* **2005**, *109*, 8755–8760. [CrossRef]
28. Tsolakis, K.P.; Kallitsis, J.K. Synthesis and Characterization of Luminescent Rod–Coil Block Copolymers by Atom Transfer Radical Polymerization: Utilization of Novel End-Functionalized Terfluorenes as Macroinitiators. *Chem. Eur. J.* **2003**, *9*, 936–943. [CrossRef]
29. Gong, X.; Iyer, P.K.; Moses, D.; Bazan, G.C.; Heeger, A.J.; Xiao, S.S. Stabilized Blue Emission from Polyfluorene-Based Light-Emitting Diodes: Elimination of Fluorenone Defects. *Adv. Funct. Mater.* **2003**, *13*, 325–330. [CrossRef]
30. Andrikopoulos, K.; Anastasopoulos, C.; Kallitsis, J.K.; Andreopoulou, A.K. Bis-Tridendate Ir(III) Polymer-Metallocomplexes: Hybrid, Main-Chain Polymer Phosphors for Orange–Red Light Emission. *Polymers* **2020**, *12*, 2976. [CrossRef]
31. Andrikopoulos, K.C.; Tselekidou, D.; Anastasopoulos, C.; Papadopoulos, K.; Kyriazopoulos, V.; Logothetidis, S.; Kallitsis, J.K.; Gioti, M.; Andreopoulou, A.K. Fluorescent Aromatic Polyether Sulfones: Processable, Scalable, Efficient, and Stable Polymer Emitters and Their Single-Layer Polymer Light-Emitting Diodes. *Nanomaterials* **2024**, *14*, 1246. [CrossRef]
32. Tselekidou, D.; Papadopoulos, K.; Kyriazopoulos, V.; Andrikopoulos, K.C.; Andreopoulou, A.K.; Kallitsis, J.K.; Laskarakis, A.; Logothetidis, S.; Gioti, M. Photophysical and Electro-Optical Properties of Copolymers Bearing Blue and Red Chromophores for Single-Layer White OLEDs. *Nanomaterials* **2021**, *11*, 2629. [CrossRef] [PubMed]
33. Jellison, G.; Modine, F. Parameterization of the optical functions of amorphous materials in the interband region. *Appl. Phys. Lett.* **1996**, *69*, 371–373. [CrossRef]
34. Azzam, R.; Bashara, N. *Ellipsometry and Polarized Light*; North-Holland Pub.: Amsterdam, The Netherlands, 1977; ISBN 0720406943.
35. Tauc, J.; Grigorovici, R.; Vancu, A. Optical Properties and Electronic Structure of Amorphous Germanium. *Phys. Status Solidi B* **1966**, *15*, 627–637. [CrossRef]
36. Gioti, M.; Tselekidou, D.; Panagiotidis, L.; Kyriazopoulos, V.; Simitzi, K.; Andreopoulou, A.K.; Kalitsis, J.K.; Gravalidis, C.; Logothetidis, S. Optical characterization of organic light-emitting diodes with selective red emission. *Mater. Today Proc.* **2021**, *37*, A39–A45. [CrossRef]
37. Oh, E.J.; Lee, J.S.; Suh, J.S.; Cho, I.H. Preparation and Photoluminescence Characteristics of Carbazole-Poly(p-phenylenevinylene)s. *Polym. J.* **2002**, *34*, 81–84. [CrossRef]
38. Li, Y.; Ding, J.; Day, M.; Tao, Y.; Lu, J.; D'iorio, M. Synthesis and Properties of Random and Alternating Fluorene/Carbazole Copolymers for Use in Blue Light-Emitting Devices. *Chem. Mater.* **2004**, *16*, 2165–2173. [CrossRef]
39. Wong, W.-Y.; Liu, L.; Cui, D.; Leung, L.M.; Kwong, C.F.; Lee, T.H.; Ng, H.-F. Synthesis and Characterization of Blue-Light-Emitting Alternating Copolymers of 9,9-Dihexylfluorene and 9-Arylcarbazole. *Macromolecules* **2005**, *38*, 4970–4976. [CrossRef]
40. Zhang, Q. Film Morphology and Electroluminescence of Poly[9,9-di-(2′-ethylhexyl) fluorenyl-2,7-diyl] Blended with a Hole Transporting Polymer. *Polym. Polym. Compos.* **2010**, *18*, 469–542. [CrossRef]
41. Bernardo, G.; Charas, A.; Morgad, J. Luminescence properties of poly(9,9-dioctylfluorene)/polyvinylcarbazole blends: Role of composition on the emission colour stability and electroluminescence efficiency. *J. Phys. Chem. Solids* **2010**, *71*, 340–345. [CrossRef]
42. Turak, A. On the Role of LiF in Organic Optoelectronics. *Electron. Mater.* **2021**, *2*, 198–221. [CrossRef]
43. Hwang, D.H.; Kim, S.K.; Park, M.-J.; Lee, J.-H.; Koo, B.-W.; Kang, I.-N.; Kim, S.-H.; Zyung, T. Conjugated Polymers Based on Phenothiazine and Fluorene in Light-Emitting Diodes and Field Effect Transistors. *Chem. Mater.* **2004**, *16*, 1298–1303. [CrossRef]
44. Park, S.H.; Kim, J.Y.; Kim, S.H.; Jin, Y.; Kim, J.; Suh, H.; Lee, K. High-efficiency new polymer light-emitting diodes with a stabilized blue emission. In *Organic Light-Emitting Materials and Devices IX*; SPIE: Bellingham, WA, USA, 2014; Volume 5937, pp. 153–158. [CrossRef]
45. Deksnys, T.; Simokaitiene, J.; Keruckas, J.; Volyniuk, D.; Bezvikonnyi, O.; Cherpak, V.; Stakhira, P.; Ivaniuka, K.; Helzhynskyy, I.; Baryshnikov, G.; et al. Synthesis and characterization of carbazole-based bipolar exciplex forming compound for efficient and colortunable OLEDs. *New J. Chem.* **2017**, *41*, 559–568. [CrossRef]

46. Xing, X.; Wu, Z.; Sun, Y.; Liu, Y.; Dong, X.; Li, S.; Wang, W. The Optimization of Hole Injection Layer in Organic Light-Emitting Diodes. *Nanomaterials* **2024**, *14*, 161. [CrossRef] [PubMed]
47. Murawski, C.; Gather, M.C. Emerging Biomedical Applications of Organic Light-Emitting Diodes. *Adv. Optical Mater.* **2021**, *9*, 2100269. [CrossRef]

Disclaimer/Publisher's Note: The statements, opinions and data contained in all publications are solely those of the individual author(s) and contributor(s) and not of MDPI and/or the editor(s). MDPI and/or the editor(s) disclaim responsibility for any injury to people or property resulting from any ideas, methods, instructions or products referred to in the content.

Article

Tunable Unexplored Luminescence in Waveguides Based on D-A-D Benzoselenadiazoles Nanofibers

Carlos Tardío [1], Esther Pinilla-Peñalver [2], Beatriz Donoso [3] and Iván Torres-Moya [4],*

[1] Department of Inorganic, Organic Chemistry and Biochemistry, Faculty of Chemical Science and Technologies, Instituto Regional de Investigación Científica Aplicada (IRICA), University of Castilla-La Mancha, 13071 Ciudad Real, Spain; carlos.tardio@uclm.es

[2] Department of Analytical Chemistry and Food Technology, University of Castilla-La Mancha, Avenue Camilo José Cela, s/n, 13071 Ciudad Real, Spain; esther.pinilla@uclm.es

[3] Department of Organic Chemistry, Faculty of Sciences, Campus of Fuentenueva, University of Granada, 18071 Granada, Spain; beatrizdonoso@ugr.es

[4] Department of Organic Chemistry, Faculty of Chemical Sciences, Campus of Espinardo, University of Murcia, 30010 Murcia, Spain

* Correspondence: ivan.torres@um.es

Abstract: A set of novel Donor-Acceptor-Donor (D-A-D) benzoselenadiazole derivatives has been synthesized and crystallized in nanocrystals in order to explore the correlation between their chemical structure and the waveguided luminescent properties. The findings reveal that all crystals exhibit luminescence and active optical waveguiding, demonstrating the ability to adjust their luminescence within a broad spectral range of 550–700 nm depending on the donor group attached to the benzoselenadiazole core. Notably, a clear relationship exists between the HOMO-LUMO energy gaps of each compound and the color emission of the corresponding optical waveguides. These outcomes affirm the feasibility of modifying the color emission of organic waveguides through suitable chemical functionalization. Importantly, this study marks the first utilization of benzoseleniadiazole derivatives for such purposes, underscoring the originality of this research. In addition, the obtention of nanocrystals is a key tool for the implementation of miniaturized photonic devices.

Keywords: benzoselenadiazoles; nanofibers; D-A-D systems; luminescence; waveguide

Citation: Tardío, C.; Pinilla-Peñalver, E.; Donoso, B.; Torres-Moya, I. Tunable Unexplored Luminescence in Waveguides Based on D-A-D Benzoselenadiazoles Nanofibers. *Nanomaterials* **2024**, *14*, 822. https://doi.org/10.3390/nano14100822

Academic Editor: Zhixing Gan

Received: 1 April 2024
Revised: 26 April 2024
Accepted: 5 May 2024
Published: 7 May 2024

Copyright: © 2024 by the authors. Licensee MDPI, Basel, Switzerland. This article is an open access article distributed under the terms and conditions of the Creative Commons Attribution (CC BY) license (https://creativecommons.org/licenses/by/4.0/).

1. Introduction

In the past decade, rapid advancements in the field of photonic devices have brought about an unprecedented transformation in optics, catalyzing breakthroughs spanning from communications to medicine [1–4]. In this dynamic landscape, optical waveguides have emerged as foundational building blocks, playing an irreplaceable role in manipulating and transmitting light at nanoscale dimensions. Their ability to guide light through microscopic structures has spearheaded a revolution in the miniaturization and efficiency of photonic systems [5–8].

Against the backdrop of this revolution, a pressing demand arises: the need for even smaller and more efficient photonic devices. Miniaturization has become a crucial pillar for developing more advanced technologies; in this context, nanoscale optical waveguides emerge as prominent players [9]. The capability of these structures to direct and modulate light at sub-microscopic levels proves crucial for adapting photonic devices to the emerging demands of the modern era. In addition, one of the main problems with organic crystals for implementation in photonic devices such as optical waveguides is their stiffness. In this sense, the search for flexible organic crystals has been a tremendously expanding line of research in recent years [10–12]. Obtaining nanostructures also contributes to the possibility of increasing their flexibility to be implemented in photonic devices, since longer structures are more difficult to deform and recover their original shape without breaking.

However, miniaturization alone is not sufficient. The precise modulation of emission color in these optical waveguides stands out as an imperative, radically transforming the functionality of these devices. From quantum communication to exploration in biophotonics, the significance of accurately tuning emission color lies not only in miniaturization, but also in the ability to customize and optimize these structures at nanoscale levels [13].

In the last decade, the benzoselenodiazole core has emerged as a chemically intriguing entity, particularly in photophysical and electronic applications. It has exhibited notable electrochemical properties, propelling its application in electronic or photovoltaic devices [14–16]. This heterocyclic system, which amalgamates a benzene ring with a selenodiazole ring, has proven exceptionally promising in the design and development of compounds with unique optical and electronic properties. Additionally, this core possesses an acceptor character, easily modulating with the introduction of donor groups at positions 4 and 7, thereby forming donor-acceptor-donor (D-A-D) systems capable of modulating HOMO (Highest Occupied Molecular Orbital) and LUMO (Lowest Unoccupied Molecular Orbital) facilitating the generation of intramolecular charge transfer (ICT) states, resulting in significant bathochromic emission shifts and achieving red-shifting properties [17–21].

The strategic presence of selenium in the structure of benzoselenodiazole imparts distinctive photophysical characteristics to these compounds, making them ideal candidates for applications in optoelectronic devices. Their ability to absorb and emit light within specific ranges of the electromagnetic spectrum has spurred intensive research in the design of photosensitive materials and efficient light-emitting compounds [22,23]. However, its application in emissive optical waveguides is still unexplored, and to the best of our knowledge, there are no examples in the literature about this application, and this fact directed us to explore the employment of this moiety in this application.

Drawing from these characteristics and our prior exploration of benzoazole-based optical waveguides [24], different benzoselenadiazole derivatives were synthesized (**1**) with the introduction of four diverse donor groups in the peripheral positions 4 and 7, connected by a π-bridge through an alkynyl group, aiming to fine-tune the emission properties and build D-π-A-π-D architectures (Figure 1). Nanocrystalline forms of these compounds were cultivated via straightforward solution-based techniques for the investigation of their luminescent and light-conducting attributes in optical waveguides.

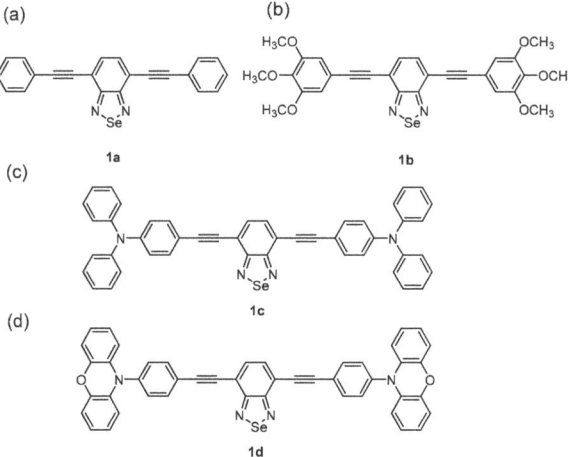

Figure 1. Chemical structures of the compounds **1a** (a), **1b** (b), **1c** (c) and **1d** (d), as described in this work.

2. Materials and Methods

2.1. General Techniques

All the chemicals utilized in the experiments for the synthesis of derivatives **1** were procured commercially, except styrene, which underwent prior distillation. Reactions involving air-sensitive materials were conducted under an atmosphere of argon. Microwave irradiations were executed using a Discover® (CEM, Matthews, NC, USA). focused microwave reactor. Silica gel, (Merck, Kieselgel 60, 230–240 mesh, Merck, Darmstadt, Germany), was employed for flash chromatography. Analytical thin layer chromatography (TLC) was carried out on aluminum-coated Merck Kieselgel 60 F254 plates (Merck, Darmstadt, Germany).

^{1}H-NMR and ^{13}C-NMR spectra collection were recorded using a Bruker Advance Neo NMR spectrometer that operates at 500.16 MHz for ^{1}H and 125.75 MHz for ^{13}C. All spectra were acquired at 298 K, employing partially deuterated solvents as internal references. Coupling constants (J) are expressed in hertz (Hz), and chemical shifts (δ) are reported in parts per million (ppm). Multiplicities are described in the following way: s = singlet, d = doublet, t = triplet, m = multiplet.

For elemental analysis experiments (EA) C, H, N and S micro-sample elemental analyzer LECO (model CNHS-932, St. Joseph, MI, USA) was used, employing 2 mg of sample for each experiment.

Mass spectra were obtained on a Bruker Autoflex II TOF/TOF (Bruker, Billerica, MA, USA) spectrometer using dithranol as the matrix for all the experiments.

UV–visible and fluorescence spectra studies in solution were carried out using a Jasco V-750 spectrophotometer (JASCO-Spain, Madrid, Spain) and Jasco FP-8300 spectrofluorimeter (JASCO-Spain, Madrid, Spain), respectively. The absorption and emission spectra were taken using dichloromethane as a solvent and at a concentration of 10^{-5} M at room temperature using standard quartz cells of 1 cm width and high spectroscopic grade solvents with very high purity.

Fluorescence images of the nanocrystals formed from derivatives **1** were captured using a Leica TCE SP2 (Wetzlar, Germany) confocal microscope equipped with a versatile mercury lamp capable of exciting fluorescence at various wavelengths. To achieve precise excitation and absorption, a set of filters was employed. These filters were designed to specifically target wavelengths within the blue spectrum (λexc = 320–380 nm, λem = 410–510 nm), green spectrum (λexc = 450–490 nm, λem = 515–565 nm) or red spectrum (λexc = 475–495 nm, λem = 520–570 nm).

2.2. Experimental Section

General procedure: A mixture of 4,7-dibromobenzo[c][1,2,5]selenadiazole (**2**) (0.100 g, 0.29 mmol), the corresponding acetylene derivative (**3**) (0.6 mmol), DBU (0.088 g, 0.58 mmol), CuI (0.003 g, 0.015 mmol) and Pd- EncatTM TPP30 (0.026 g, 0.01 mmol) was charged to a dried microwave vessel under an argon atmosphere. After that, 1 mL of CH$_3$CN was added to the vessel, which was then closed and irradiated at 150 °C for 20 min in all the cases. The crudes' reactions were purified by column chromatography using hexane/ethyl acetate as eluent to achieve analytically pure products **1**. It can be pointed out that for all the reactions, microwave irradiation was used as an energy source and Pd-EncatTM TPP30 was recovered after its use by filtration, and was used again for following reactions to reduce the environmental impact of the process.

-4,7-bis(phenylethynyl)benzo[c][1,2,5]selenadiazole (**1a**): From ethynylbenzene (**3a**) (0.061 g, 0.6 mmol), derivative **1a** (0.095 g, 86%) was obtained as a yellow solid by column chromatography using hexane/ethyl acetate as eluent (9:1). ^{1}H-NMR (CDCl$_3$, 300 MHz), d: 7.54 (s, 2H, H$_{benzoselenazole}$), 7.52–7.51 (m, 4H, H$_{phenyl}$), 7.22–7.20 (m, 6H, H$_{phenyl}$), ^{13}C-NMR (CDCl$_3$, 75 MHz), d: 155.5, 131.4, 129.6, 128.7, 128.5, 123.3, 110.0, 91.9, 75.6. MS (EI): m/z: 383.9 [M+]. EA; Calculated for C$_{22}$H$_{12}$N$_2$Se; C: 68.94; H: 3.16; N: 7.31, found C: 68.97; H: 3.13; N: 7.28.

-4,7-bis((3,4,5-trimethoxyphenyl)ethynyl)benzo[c][1,2,5]selenadiazole (**1b**): From 5-ethynyl-1,2,3-trimethoxybenzene (**3b**) (0.115 g, 0.6 mmol), derivative **1b** (0.148 g, 91%) was obtained as a dark yellow solid by column chromatography using hexane/ethyl acetate (9:1) as eluent. ^{1}H-NMR (CDCl$_3$, 300 MHz), δ: 7.51 (s, 2H, H$_{benzoselenazole}$), 6.86 (s, 4H, H$_{phenyl}$), 3.82 (s, 18H, -OCH$_3$), ^{13}C-NMR (CDCl$_3$, 75 MHz), d: 155.5, 154.8, 139.0, 129.6, 118.3, 110.0, 106.4, 95.0, 74.2, 60.7, 56.8. MS (EI): m/z: 565.41 [M+]. EA; Calculated for C$_{28}$H$_{24}$N$_2$O$_6$Se; C: 59.68; H: 4.29; N: 4.97, found C: 59.70; H: 4.32; N: 5.00.

-4,7-bis(diylbis(ethyne-2,1-diyl))bis(N,N-diphenylaniline)benzo[c][1,2,5]selenadiazole (**1c**) From 4-ethynyl-N,N-diphenylaniline (**3c**) (0.161 g, 0.6 mmol), derivative **1c** (0.162 g, 78%) was obtained as an orange solid by column chromatography using hexane/ethyl acetate (9:1) as eluent. ^{1}H-NMR (CDCl$_3$, 300 MHz), δ: 7.58 (s, 2H, H$_{benzoselenazole}$), 7.49–7.47 (d, 4H, H$_{phenyl}$), 7.26–7.23 (t, 8H, H$_{phenyl}$), 7.09–7.07 (m, 12H, H$_{phenyl}$), 6.98–6.94 (m, 4H, H$_{phenyl}$). ^{13}C-NMR (CDCl$_3$, 75 MHz), d: 155.5, 145.6, 145.4, 131.4, 129.8, 129.6, 128.3, 126.3, 124.9, 119.7, 110.0, 91.9, 75.6. MS (EI): m/z: 719.1 [M+]. EA; Calculated for C$_{46}$H$_{30}$N$_4$Se; C: 76.98, H: 4.21; N: 7.81, found C: 76.99, H: 4.22; N: 7.80.

-4,7-bis((4-(10H-phenoxazin-10-yl)phenyl)ethynyl)benzo[c][1,2,5]selenadiazole (**1d**). From 10-(4-ethynylphenyl)-10H-phenoxazine (**3d**) (0.170 g, 0.6 mmol), derivative **1d** (0.140 g, 65%) was obtained as a red solid by column chromatography using hexane/ethyl acetate (95:5) as eluent. ^{1}H-NMR (CDCl$_3$, 300 MHz), δ: 7.52–7.51 (d, 4H, H$_{phenyl}$), 7.44 (s, 2H, H$_{benzoselenazole}$), 7.32–7.30 (m, 4H, H$_{phenyl}$), 7.24–7.18 (m, 12H, H$_{phenyl}$), 7.16–7.15 (d, 4H, H$_{phenyl}$). ^{13}C-NMR (CDCl$_3$, 75 MHz), δ: 155.5, 145.7, 144.9, 132.9, 131.5, 130.3, 129.6, 125.5, 125.2, 120.8, 119.4, 117.4, 110.0, 91.9, 75.6. MS (EI): m/z: 746.1 [M+]. EA; Calculated for C$_{46}$H$_{26}$N$_4$O$_2$Se; C: 74.09, H: 3.51; N: 7.51, O: 4.29, found C: 74.13, H: 3.49; N: 7.52, O: 4.29.

3. Results and Discussion

3.1. Synthesis

The synthesis of D-A-D benzoselenadiazole derivatives **1** was performed through a Sonogashira C-C cross-coupling reaction between derivatives **2** and **3** (Scheme 1), optimized previously by our research group for other azoles and benzoazoles [25,26]. All the reactions were carried out using microwave irradiation, with the reusable catalyst Pd-Encat TPP30 to increase the sustainability of the synthetic procedure, achieving the desired compounds **1** in 30 min in very good yields (65–91%), that gave adequate analytical NMR spectroscopic data (NMR spectra collection is recorded in Supporting Information section). We would like to point out that derivative **1a** had been previously synthesized by Li and co-workers under conventional conditions [27], improving the yield with this procedure using microwave irradiation as an energy source and considerably reducing the reaction time.

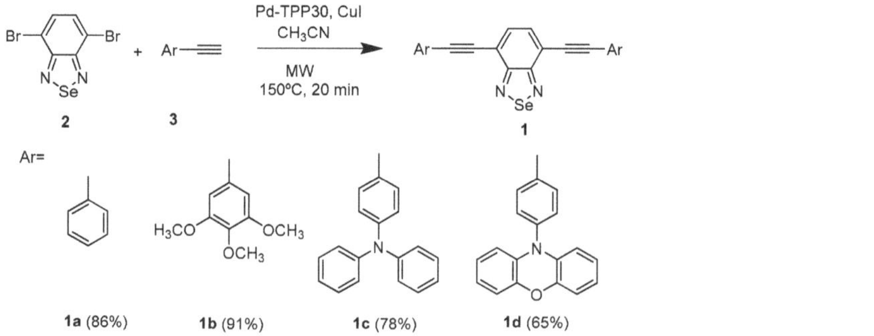

Scheme 1. Synthesis of D-A-D benzoselenadiazoles derivatives **1**, as described in this work.

3.2. Theoretical Calculations

Before later optical characterization, the minimum-energy optimized structures were calculated at the B3LYP/6-31G (d,p) theoretical level [28,29] and are detailed in Table 1.

Across all compounds, the HOMO molecular orbital is situated along the horizontal axis of the molecule, mainly affected by the donor group, while the LUMO orbital predominantly occupies the vertical axis, which mainly corresponds to the benzoselenadiazole core, existing as a shared overlap region, thus facilitating intramolecular charge transfer (ICT) [30].

Table 1. Energies and topologies of the HOMO and LUMO frontier molecular orbitals, and HOMO-LUMO gaps of benzoselenadiazoles 1 obtained using DFT calculations at the B3LYP/6-31G (d,p) theory level.

Compound	HOMO (eV)	LUMO (eV)	HOMO LUMO Gap (eV)
1a	−5.66	−3.06	2.60
1b	−5.58	−3.14	2.44
1c	−5.28	−3.08	2.20
1d	−5.11	−3.06	2.05

The variation in donor peripheral groups (1a–d) correlates with the electron-donating tendency of the substituent, as indicated by the HOMO-LUMO gap. The most dramatical changes can be detected in HOMO values due to the changes in the donor group that implies changes in the HOMO-LUMO gaps, while LUMO values are not significantly affected because the major contribution is because the benzoselenadiazole moiety that keeps constant in all the derivatives. Specifically, compound 1a, featuring peripheral phenyl groups (weaker donor groups), exhibits a higher HOMO-LUMO gap value compared to compounds 1b–d. The introduction of methoxy groups in the periphery in 1b with +I and +K characters increases the donor character and decreases the HOMO-LUMO gap with respect to 1a (2.44 vs. 2.60 eV). Consequently, 1c, with enhanced donor character due to its triphenylamine group, demonstrates a lower HOMO-LUMO gap value relative to 1a and 1b (2.20 vs. 2.60 and 2.44 eV). Finally, 1d, with the strongest donor group, phenylphenoxazine, showed the lowest HOMO-LUMO gap (2.05 eV), and this fact will directly impact their photophysical properties.

3.3. Photophysical Studies

The UV-vis and fluorescence spectra of the different derivatives 1 were carried out in 10^{-5} M CH_2Cl_2 solutions. They are recorded in Figure 2 and photophysical data are summarized in Table 2.

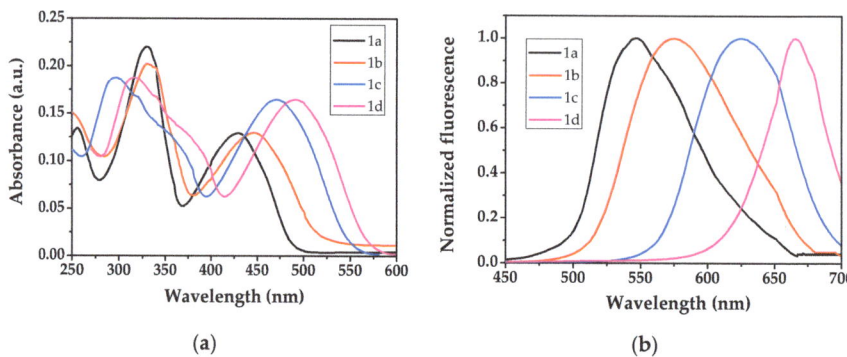

Figure 2. Absorption (a) and emission (b) spectra of derivatives **1** in 10^{-5} M CH_2Cl_2 solutions.

Table 2. Photophysical data of derivatives **1**.

Compound	λ_{abs} [a] (nm)	λ_{em} [b] (nm)	λ_{emsol} [c] (nm)	λ_{onset} [d] (nm)	Φ_{dis} [e]	Φ_{sol} [f]	HOMO-LUMO Gap [g] (eV)	HOMO-LUMO Gap [h] (eV)
1a	329, 434	546	550	488	0.20	0.13	2.60	2.54
1b	330, 448	574	595	514	0.54	0.43	2.44	2.41
1c	300, 470	625	640	552	0.69	0.41	2.20	2.25
1d	315, 491	664	698	585	0.60	0.35	2.05	2.11

[a] Absorption maxima of derivatives **1** in solution; [b] Emission maxima of derivatives **1** in solution; [c] Emission maxima of nanocrystals of derivatives **1**; [d] Onset wavelength calculated from the absorption spectra of the lowest energy band; [e] PL quantum yield in solution of derivatives **1**. [f] PL quantum yield measured in the solid state; [g] Theoretical HOMO-LUMO gap of compounds **1** computed at B3LYP/6-31G(d,p) level. [h] Experimental HOMO-LUMO gap of compound **1** calculated from the onset of the lowest energy absorption band using the $1240/\lambda_{onset}$ formula.

It can be pointed out that the data recorded in Table 2 for **1a** agree with the data previously described by Li and co-workers for this compound [27]. The compounds **1a–d** exhibit absorption spectra characterized by a broad band with peaks ranging from around 300 to 330 nm, attributed to a π-π* transition, and a second band centered at longer wavelengths (around 430–490 nm), attributed to an ICT state. When the photoluminescence (PL) spectra are calculated following photoexcitation at the peaks of the longest-wavelength absorption band, the maxima shifts from 546 nm to 664 nm. This gradual red-shift of the PL spectra corresponds to the push–pull nature of the dyes, moving towards longer wavelengths as the donor-acceptor character strengthens. Consequently, **1d** displays the longest emission wavelength among all D-A-D derivatives due to the pronounced electron-donating properties of the phenylphenoxazine moiety. Similarly, **1c** also exhibits bathochromic shifts compared to **1a** and **1b**, attributed to the enhanced donor character of the triphenylamine derivative in comparison to phenyl or trimethoxyphenyl groups.

Consequently, the photoluminescence (PL) spectra of the derivatives **1** undergo spectral changes in alignment with the calculated HOMO-LUMO gap energies. The experimental optical HOMO-LUMO gaps were determined based on the onset of the lowest energy band absorption and demonstrated satisfactory agreement with theoretical predictions (Table 2). Additionally, it is noteworthy that a similar observation can be made regarding the spectral evolution of the lowest π-π* absorption transition..

Finally, the fluorescence quantum yields of these derivatives in solution were determined in $CHCl_3$ using quinine sulfate in 0.1 M H_2SO_4 (Φ = 0.54) and fluorescein in 0.1 M NaOH (Φ = 0.79) for **1a–c**, and cresyl violet in ethanol (Φ = 0.56) for **1d** as standards, revealing moderate values (Table 2). The values of the quantum yield in the solid state

(Table 2) are moderate and slightly decreased for all the derivatives in comparison with values in the solution state to values between 0.13 and 0.43.

3.4. Self-Assembling Studies

The creation of supramolecular assemblies using derivatives **1** was achieved through the slow diffusion method. In this procedure, a container holding a diluted solution (1 mg) of derivatives **1** in 1 mL of either chloroform or tetrahydrofuran, recognized as effective solvents, was carefully placed into another container containing a non-soluble poor solvent such as hexane, acetonitrile, ethanol or methanol. The second container was securely sealed and left undisturbed at room temperature. After a few days, the emergence of supramolecular aggregates becomes observable.

SEM images of the resultant nanofibers were acquired, with a focus on selecting the most promising ones in the form of fibers. Examples showcasing the optimal morphologies are presented in Figure 3, while additional SEM images can be found in the Supporting Information (Figure S1).

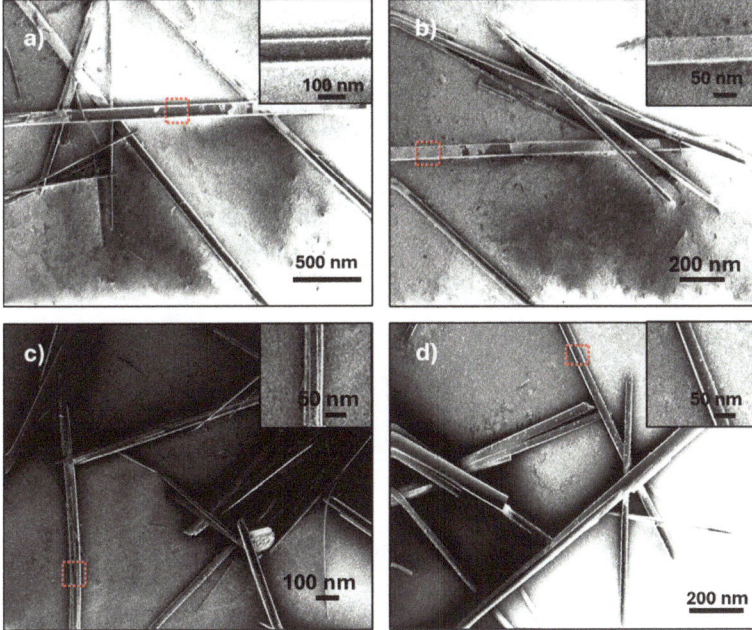

Figure 3. SEM images for the nanocrystals of (**a**) **1a** (THF/MeOH); (**b**) **1b** (CHCl$_3$/EtOH); (**c**) **1c** (CHCl$_3$/MeOH); (**d**) **1d** (THF/EtOH).

As we can observe in Figure 3, all nanocrystals exhibit a well-defined fibrillar shape, making them ideal candidates for investigating their optical waveguide properties. It can be pointed out that the best fibrillar morphologies were obtained using MeOH or EtOH as poor solvents.

3.5. Optical Waveguiding Behaviour

After identifying the most promising crystalline microfibers with suitable morphologies, confocal fluorescence microscopy images (Figure 4) of these nanocrystals were taken and recorded to assess their optical waveguiding characteristics.

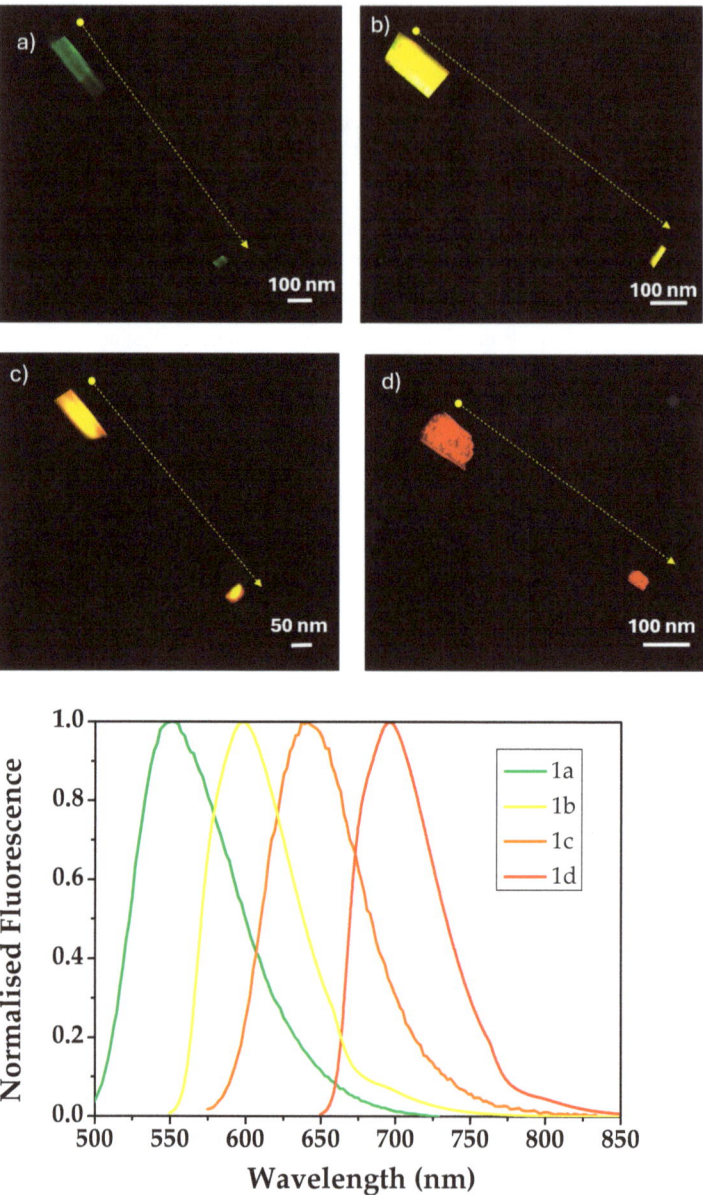

Figure 4. Confocal microscopy images and photoluminescence spectra for the nanocrystals of **1a** (a), **1b** (b), **1c** (c) and **1d** (d).

The outcomes revealed that all nanocrystals of compounds **1a–d** exhibited luminescence. Upon irradiation with a 365 nm laser at the fiber body, it was notable that the emitted light propagated to the crystal's tip, while the body itself brightened with emission. This phenomenon indicated strong luminescence with a distinct contrast between the bright tip/edges and the darker bulk, confirming efficient photon transport along the nanocrystals. Additionally, each nanocrystal was emitted at varying wavelengths, all distinct from the excitation wavelength, thus affirming their active optical waveguiding behavior (Figure 4). In addition, we can observe that in function on the molecular structure and the substitution in the benzoselenadiazole moiety, nanocrystals emitted a broad range of wavelengths, spanning from 550–700 nm. Notably, in the solid state, a significant redshift in PL is observed compared to the solution phase (Table 2 and Figure 2b vs. Figure 4). It is important to highlight that structural modifications of these derivatives exert an identical influence in both solution and solid states, resulting in consistent bathochromic shifts as the donor-acceptor (D-A) character of the derivative intensifies. Comparison of the emission wavelengths with theoretically calculated HOMO-LUMO gap values at the B3LYP/6-31G (d,p) theory level revealed a clear correlation, demonstrating the influence of different functional groups on the optical waveguide emission color. Fluorescence microscopy images revealed a progressive red-shift in the emission from green to red as theoretically calculated HOMO-LUMO gap energies decreased, indicating that TD-DFT could be utilized for on-demand emission tuning of the optical waveguides, and confirmed by the experimental HOMO-LUMO gaps obtained through the onset in absorption spectra. Like this, beginning with the highest HOMO-LUMO gap (2.60 eV), compound **1a** exhibited green emission (550 nm), compound **1b**, with the introduction of the trimethoxy groups, induces a lower HOMO-LUMO gap (2.44 eV), emitting in yellow (595 nm). The introduction of a stronger donor group like triphenylamine decreases the HOMO-LUMO gap (2.20 eV), achieving an orange color emission in the nanocrystals (640 nm). Finally, **1d**, with the strongest donor group (phenylphenoxazine), shows the lowest HOMO-LUMO gap (2.05 eV), with strong emission in red in the nanocrystals (698 nm). (Figure 5).

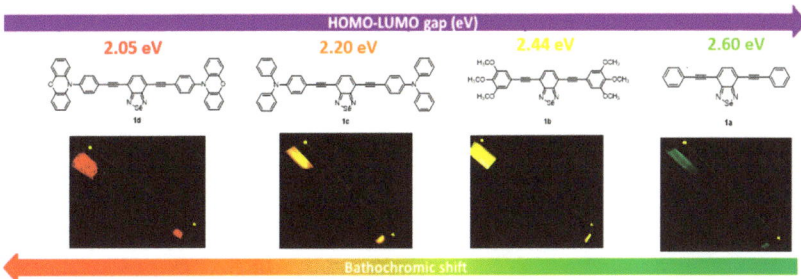

Figure 5. Theoretical HOMO-LUMO gap of derivatives **1a–d**/emission color relationship of the optical waveguides obtained from the nanocrystals of benzoselenadiazoles **1**.

While numerous studies have explored this subject, as far as we know, there is currently no literature correlating the variation of emission color of a waveguide for benzoseleniadiazole derivatives, highlighting the novelty of this work.

Moreover, it is noteworthy that derivatives **1a–d** exhibit a tendency to form cross-linked and bunch-shaped nanocrystals as SEM images reveal. Irradiation of a single point on one fiber facilitated the light propagation in different directions; for example, in **1a** (Figure 6). This curious behavior, as described in recent literature [31–33], enhances the practical utility of these materials in real-world devices. It enables light to travel through different channels, facilitating parallel connectivity and interconnection between nanowires and various devices and facilitating the implementation in real miniaturized devices, which are very in demand these days in current society. For all these reasons, these results may be very useful in the coming years for the development of miniaturized photonic chips.

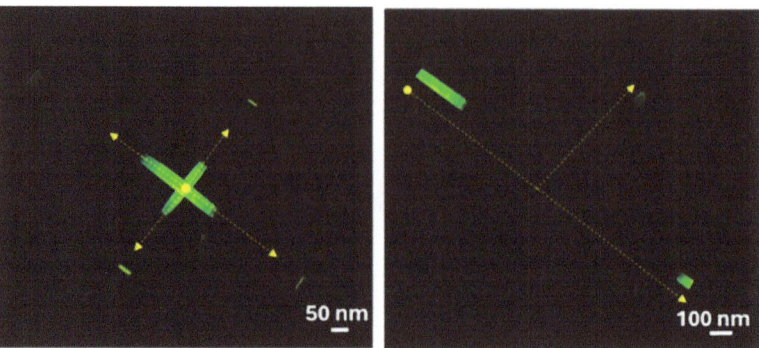

Figure 6. Confocal fluorescence microscopy images for interconnected fibers of **1a** showing the light transmission for different points in different directions.

4. Conclusions

This study examines the impact of various peripheral donor groups incorporated into the benzoselenadiazole core, resulting in the design of diverse D-A-D derivatives. The investigation focuses on their performance as waveguides and their effect on emission color.

Nanocrystals formed via the self-assembly of the benzoselenadiazoles **1a–d** using the slow diffusion technique were examined through SEM images. The findings illustrate a significant inclination towards nanofiber formation, making them suitable for studies on optical waveguiding.

Each of the crystals of compounds **1a–d** exhibited efficient optical waveguiding behavior, accompanied by tunable color emissions ranging from green to red, contingent upon the nature of the peripheral donor groups. Notably, a clear correlation between the HOMO-LUMO gap and the emitted color of the optical waveguide exists. A reduced HOMO-LUMO gap resulted in a shift towards longer wavelengths in the emission spectrum. Compound **1d** shows the lowest HOMO-LUMO gap, and as a consequence, the most red-shifted emission.

This successful discovery could serve as a valuable approach for adjusting the emission color of optical waveguides, offering a systematic method for designing organic compounds and exploring their potential utilization as organic waveguides. Finally, the nanoscopic nature of these nanofibers and the properties described in this work constitute a promising study for the implementation of these nanofibers in real photonic devices, which are in high demand in current society.

Supplementary Materials: The following supporting information can be downloaded at: https://www.mdpi.com/article/10.3390/nano14100822/s1, Figure S1: SEM images; Figures S2–S9: NMR spectra.

Author Contributions: Conceptualization, I.T.-M.; methodology, C.T., B.D. and E.P.-P.; software, I.T.-M.; validation, I.T.-M.; formal analysis, C.T. and E.P.-P.; investigation, I.T.-M., B.D., C.T. and E.P.-P.; resources, I.T.-M.; data curation, C.T., B.D. and E.P.-P.; writing—original draft preparation, I.T.-M., B.D., C.T. and E.P.-P.; writing—review and editing, I.T.-M., B.D., C.T. and E.P.-P.; visualization, I.T.-M.; supervision, I.T.-M.; project administration, I.T.-M.; funding acquisition, I.T.-M. All authors have read and agreed to the published version of the manuscript.

Funding: This research was funded by Juan de la Cierva Formación 2020 FJC2020-043684-I with assistance financed by MCIN/AEI/10.13039/501100011033 and for the Unión Europea NextGeneration EU/PRTR.

Data Availability Statement: Data are contained within the article.

Acknowledgments: I. Torres-Moya is indebted to Juan de la Cierva Formación 2020 FJC2020-043684-I. Computer resources, technical expertise and assistance provided by High Performance Computing Service of the University of Castilla-La Mancha was also acknowledged.

Conflicts of Interest: The authors declare no conflicts of interest.

References

1. Annadhasan, M.; Karothu, D.P.; Chinnasamy, R.; Catalano, L.; Ahmed, E.; Ghosh, S.; Naumov, P.; Chandrasekar, R. Micromanipulation of mechanically compliant organic single-crystal optical microwaveguides. *Angew. Chem. Int. Ed.* **2020**, *59*, 13821–13830. [CrossRef] [PubMed]
2. Jamil, B.; Choi, Y. Soft Optical Waveguide Sensors Tuned by Reflective Pigmentation for Robotic Applications. *J. Korea Robot. Soc.* **2021**, *16*, 1–11. [CrossRef]
3. Butt, M.A.; Khonina, S.N.; Kazanskiy, N.L. Recent Advances in Photonic Crystal Optical Devices: A Review. *Opt. Laser Technol.* **2021**, *142*, 107265. [CrossRef]
4. Shahbaz, M.; Butt, M.A.; Piramidowicz, R. A Concise Review of the Progress in Photonic Sensing Devices. *Photonics* **2023**, *10*, 698. [CrossRef]
5. Chen, S.; Zhuo, M.-P.; Wang, X.-D.; Wei, G.-Q.; Liao, L.-S. Optical Waveguides Based on One-Dimensional Organic Crystals. *Photonix* **2021**, *2*, 2. [CrossRef]
6. Wu, S.; Zhou, B.; Yan, D. Recent Advances on Molecular Crystalline Luminescent Materials for Optical Waveguides. *Adv. Opt. Mater.* **2021**, *9*, 2001768. [CrossRef]
7. Tian, D.; Chen, Y. Optical Waveguides in Organic Crystals of Polycyclic Arenes. *Adv. Opt. Mater.* **2021**, *9*, 2002264. [CrossRef]
8. Shi, Y.-L.; Wang, X.-D. Organic Micro/Nanostructures for Photonics. *Adv. Funct. Mater.* **2021**, *31*, 2008149. [CrossRef]
9. Lucas, N. Integrated Photonics: Advances in Miniaturization and Integration. *J. Laser Opt. Photonics* **2023**, *10*, 6.
10. Chandrasekar, R. Mechanophotonics—A Guide to Integrating Microcrystals toward Monolithic and Hybrid All-Organic Photonic Circuits. *Chem. Commun.* **2022**, *58*, 3415–3428. [CrossRef]
11. Annadhasan, M.; Basak, S.; Chandrasekhar, N.; Chandrasekar, R. Next-generation Organic Photonics: The Emergence of Flexible Crystal Optical Waveguides. *Adv. Opt. Mater.* **2020**, *8*, 2000959. [CrossRef]
12. Ravi, J.; Chandrasekar, R. Micromechanical Fabrication of Resonator Waveguides Integrated Four-port Photonic Circuit from Flexible Organic Single Crystals. *Adv. Opt. Mater.* **2021**, *9*, 2100550. [CrossRef]
13. Di, Q.; Li, L.; Miao, X.; Lan, L.; Yu, X.; Liu, B.; Yi, Y.; Naumov, P.; Zhang, H. Fluorescence-Based Thermal Sensing with Elastic Organic Crystals. *Nat. Commun.* **2022**, *13*, 5280. [CrossRef] [PubMed]
14. Abdullah; Akhtar, M.S.; Kim, E.-B.; Fijahi, L.; Shin, H.-S.; Ameen, S. A Symmetric Benzoselenadiazole Based D–A–D Small Molecule for Solution Processed Bulk-Heterojunction Organic Solar Cells. *J. Ind. Eng. Chem.* **2020**, *81*, 309–316. [CrossRef]
15. Ting, H.-C.; Chen, Y.-H.; Lin, L.-Y.; Chou, S.-H.; Liu, Y.-H.; Lin, H.-W.; Wong, K.-T. Benzochalcogenodiazole-based Donor–Acceptor–Acceptor Molecular Donors for Organic Solar Cells. *ChemSusChem* **2014**, *7*, 457–465. [CrossRef]
16. Mondal, S.; Rashid, A.; Ghosh, P. Benzoselenadiazole Containing Donor–Acceptor–Donor Receptor as a Superior and Selective Probe for Fluoride in DMSO. *Inorganica Chim. Acta* **2022**, *538*, 120973. [CrossRef]
17. Lu, X.; Fan, S.; Wu, J.; Jia, X.; Wang, Z.-S.; Zhou, G. Controlling the Charge Transfer in D–A–D Chromophores Based on Pyrazine Derivatives. *J. Org. Chem.* **2014**, *79*, 6480–6489. [CrossRef] [PubMed]
18. Qian, G.; Dai, B.; Luo, M.; Yu, D.; Zhan, J.; Zhang, Z.; Ma, D.; Wang, Z.Y. Band Gap Tunable, Donor−Acceptor−Donor Charge-Transfer Heteroquinoid-Based Chromophores: Near Infrared Photoluminescence and Electroluminescence. *Chem. Mater.* **2008**, *20*, 6208–6216. [CrossRef]
19. Tao, T.; Ma, B.-B.; Peng, Y.-X.; Wang, X.-X.; Huang, W.; You, X.-Z. Asymmetrical/Symmetrical D−π−A/D−π−D Thiazole-Containing Aromatic Heterocyclic Fluorescent Compounds Having the Same Triphenylamino Chromophores. *J. Org. Chem.* **2013**, *78*, 8669–8679. [CrossRef]
20. Wang, J.-L.; Xiao, Q.; Pei, J. Benzothiadiazole-Based D−π−A−π−D Organic Dyes with Tunable Band Gap: Synthesis and Photophysical Properties. *Org. Lett.* **2010**, *12*, 4164–4167. [CrossRef]
21. Homnick, P.J.; Tinkham, J.S.; Devaughn, R.; Lahti, P.M. Engineering Frontier Energy Levels in Donor–Acceptor Fluoren-9-Ylidene Malononitriles versus Fluorenones. *J. Phys. Chem. A* **2014**, *118*, 475–486. [CrossRef]
22. Liang, A.; Wang, H.; Chen, Y.; Zheng, X.; Cao, T.; Yang, X.; Cai, P.; Wang, Z.; Zhang, X.; Huang, F. Benzoselenadiazole-Based Donor-Acceptor Small Molecule: Synthesis, Aggregation-Induced Emission and Electroluminescence. *Dyes Pigm.* **2018**, *149*, 399–406. [CrossRef]
23. Gao, S.; Balan, B.; Yoosaf, K.; Monti, F.; Bandini, E.; Barbieri, A.; Armaroli, N. Highly Efficient Luminescent Solar Concentrators Based on Benzoheterodiazole Dyes with Large Stokes Shifts. *Chemistry* **2020**, *26*, 11013–11023. [CrossRef]
24. Torres-Moya, I.; Martín, R.; Díaz-Ortiz, Á.; Prieto, P.; Carrillo, J.R. Self-assembled Alkynyl Azoles and Benzoazoles as Colored Optical Waveguides. *Isr. J. Chem.* **2018**, *58*, 827–836. [CrossRef]
25. Cáceres, D.; Cebrián, C.; Rodríguez, A.M.; Carrillo, J.R.; Díaz-Ortiz, Á.; Prieto, P.; Aparicio, F.; García, F.; Sánchez, L. Optical Waveguides from 4-Aryl-4H-1,2,4-Triazole-Based Supramolecular Structures. *Chem. Commun.* **2013**, *49*, 621–623. [CrossRef]

26. Pastor, M.J.; Torres, I.; Cebrián, C.; Carrillo, J.R.; Díaz-Ortiz, Á.; Matesanz, E.; Buendía, J.; García, F.; Barberá, J.; Prieto, P.; et al. 4-aryl-3,5-bis(Arylethynyl)Aryl-4H-1,2,4-triazoles: Multitasking Skeleton as a Self-assembling Unit. *Chemistry* **2015**, *21*, 1795–1802. [CrossRef]
27. Li, H.; Guo, Y.; Lei, Y.; Gao, W.; Liu, M.; Chen, J.; Hu, Y.; Huang, X.; Wu, H. D-π-A Benzo[c][1,2,5]Selenadiazole-Based Derivatives via an Ethynyl Bridge: Photophysical Properties, Solvatochromism and Applications as Fluorescent Sensors. *Dyes Pigm.* **2015**, *112*, 105–115. [CrossRef]
28. Zhao, Y.; Truhlar, D.G. Density Functionals for Noncovalent Interaction Energies of Biological Importance. *J. Chem. Theory Comput.* **2007**, *3*, 289–300. [CrossRef]
29. Francl, M.M.; Pietro, W.J.; Hehre, W.J.; Binkley, J.S.; Gordon, M.S.; DeFrees, D.J.; Pople, J.A. Self-Consistent Molecular Orbital Methods. XXIII. A Polarization-Type Basis Set for Second-Row Elements. *J. Chem. Phys.* **1982**, *77*, 3654–3665. [CrossRef]
30. Guo, Z.-H.; Lei, T.; Jin, Z.-X.; Wang, J.-Y.; Pei, J. T-Shaped Donor–Acceptor Molecules for Low-Loss Red-Emission Optical Waveguide. *Org. Lett.* **2013**, *15*, 3530–3533. [CrossRef]
31. Zheng, J.Y.; Yan, Y.; Wang, X.; Zhao, Y.S.; Huang, J.; Yao, J. Wire-on-Wire Growth of Fluorescent Organic Heterojunctions. *J. Am. Chem. Soc.* **2012**, *134*, 2880–2883. [CrossRef]
32. Kong, Q.; Liao, Q.; Xu, Z.; Wang, X.; Yao, J.; Fu, H. Epitaxial Self-Assembly of Binary Molecular Components into Branched Nanowire Heterostructures for Photonic Applications. *J. Am. Chem. Soc.* **2014**, *136*, 2382–2388. [CrossRef]
33. Li, Z.-Z.; Tao, Y.-C.; Wang, X.-D.; Liao, L.-S. Organic Nanophotonics: Self-Assembled Single-Crystalline Homo-/Heterostructures for Optical Waveguides. *ACS Photonics* **2018**, *5*, 3763–3771. [CrossRef]

Disclaimer/Publisher's Note: The statements, opinions and data contained in all publications are solely those of the individual author(s) and contributor(s) and not of MDPI and/or the editor(s). MDPI and/or the editor(s) disclaim responsibility for any injury to people or property resulting from any ideas, methods, instructions or products referred to in the content.

Article

The Enhanced Photoluminescence Properties of Carbon Dots Derived from Glucose: The Effect of Natural Oxidation

Pei Zhang [1], Yibo Zheng [1], Linjiao Ren [1,*], Shaojun Li [1], Ming Feng [1], Qingfang Zhang [1], Rubin Qi [1], Zirui Qin [1], Jitao Zhang [1] and Liying Jiang [2,*]

[1] Henan Key Lab of Information-Based Electrical Appliances, College of Electrical and Information Engineering, Zhengzhou University of Light Industry, Zhengzhou 450002, China; zhangpei@zzuli.edu.cn (P.Z.); zhengyb_2018@163.com (Y.Z.); shaojunli1997@163.com (S.L.); fengming9851@163.com (M.F.); qingfang@zzuli.edu.cn (Q.Z.); qirubin@zzuli.edu.cn (R.Q.); 2019005@zzuli.edu.cn (Z.Q.); zhangjitao@zzuli.edu.cn (J.Z.)

[2] School of Electronics and Information, Academy for Quantum Science and Technology, Zhengzhou University of Light Industry, No. 136 Ke Xue Avenue, Zhengzhou 450002, China

* Correspondence: renlinjiao@zzuli.edu.cn (L.R.); jiangliying@zzuli.edu.cn (L.J.)

Abstract: The investigation of the fluorescence mechanism of carbon dots (CDs) has attracted significant attention, particularly the role of the oxygen-containing groups. Dual-CDs exhibiting blue and green emissions are synthesized from glucose via a simple ultrasonic treatment, and the oxidation degree of the CDs is softly modified through a slow natural oxidation approach, which is in stark contrast to that aggressively altering CDs' surface configurations through chemical oxidation methods. It is interesting to find that the intensity of the blue fluorescence gradually increases, eventually becoming the dominant emission after prolonging the oxidation periods, with the quantum yield (QY) of the CDs being enhanced from ~0.61% to ~4.26%. Combining the microstructure characterizations, optical measurements, and ultrafiltration experiments, we hypothesize that the blue emission could be ascribed to the surface states induced by the C–O and C=O groups, while the green luminescence may originate from the deep energy levels associated with the O–C=O groups. The distinct emission states and energy distributions could result in the blue and the green luminescence exhibiting distinct excitation and emission behaviors. Our findings could provide new insights into the fluorescence mechanism of CDs.

Keywords: carbon dots; dual-fluorescent; natural oxidation; oxygen-containing groups; fluorescence mechanism

1. Introduction

As a new type of fluorescent nanomaterials, carbon dots (CDs) have garnered increasing attention over the past two decades due to their numerous advantages, including their multi-color emission, excellent biocompatibility, high quantum yield (QY) and wide luminescence range. These properties enable their extensive applications in various potential fields such as bioimaging [1], sensors [2,3], light-emitting diodes (LEDs) [4,5], luminescent solar concentrators [6,7] and lasers [8,9]. Although the photoluminescence (PL) properties of CD materials have been extensively explored, a unified fluorescence mechanism remains elusive. This is due to the multiple emission centers and diverse electronic structures induced by the different particle sizes, conjugation lengths, surfaces, passivation, etc. [10,11]. Notably, the role of oxygen-containing groups bonded to the CD surface in fluorescence emission is a significant factor, as the oxygen atoms are readily incorporated into the CD structure during the preparation process [10,11].

The effects of oxygen-containing groups on the luminescence efficiency and wavelength of CDs have been intensively explored [12–19]. Several studies have found that the QY of CDs can be substantially enhanced by removing oxygen-related groups through violent reduction reaction treatments [12–14]. However, contrary findings have also been reported by other researchers [15–17]. Additionally, the emission wavelength of CDs tends to redshift with an increase in oxygen-containing groups, as the optical bandgap of CDs narrows after incorporating these species [18,19]. During the oxidation or reduction process, the reagents used not only alter the carbon cores but also significantly modify the surface configurations of CDs, further complicating the luminescence mechanism.

Therefore, it is essential to comprehensively understand the effect of oxygen-related surface states on the luminescence mechanisms of CDs with multiple emissions, which is crucial for tailoring their surface chemistry. In this work, dual-fluorescent (blue and green) CDs are synthesized using a one-step ultrasonication method. The surface configurations of the CDs are gently modified through aerial oxidation, resulting in an enhancement of the QY from about 0.61% to 4.26% as the oxidation degree increases. Interestingly, the intensity of both blue and green emissions shows similar variation trends with the amounts of their respective oxygenous groups during the oxidation process. Based on microstructure characterizations, optical measurements, and ultrafiltration experiments, we hypothesize that the blue emission, which exhibits excitation-independent characteristics, can be attributed to C–O and C=O groups, while the green luminescence possessing the excitation-dependent feature is likely associated with O–C=O groups. Furthermore, the emission states induced by different oxygenous groups have distinct distribution and energy levels within the bandgap of small carbonaceous materials or CDs, leading to the blue and the green luminescence exhibiting different excitation-emission properties. The enhanced luminescence efficiency can be explained by two factors: the reduction in the number of oxygen-centered radicals, as evidenced by the electron paramagnetic resonance (EPR) spectra, and the increase in the density of oxygen-related emission states.

2. Materials and Methods

2.1. The Synthesis of Dual-Fluorescent CDs

All the regents used in this experiment were purchased from Sigma Aldrich (Shanghai, China), including glucose, NaOH (97%), ethanol, HCl (37%), and $MgSO_4$ (97%). The CDs were synthesized through the following steps: (1) Mix the aqueous glucose solution (1 mol/L, 10 mL) with the NaOH solution (1.5 mol/L, 10 mL), then treat the mixture using the ultrasonication method for 4 h. The ultrasonic power and frequency were set at 400 W and 50 Hz, respectively. (2) After the ultrasonic treatment, add HCl to adjust the pH of the mixture to 7. (3) Gradually add 100 mL of ethanol using a piston burette while stirring. Subsequently, add about 17 g of $MgSO_4$ (wt%) to the mixture. The stirring speed and time were set at 2000 r/min and 20 min, respectively. (4) Finally, obtain the CDs in ethanol/water by removing the precipitate [20].

After synthesizing the CDs, we transferred 60 mL of the CD solution into a 125 mL wide-mouth bottle and placed it in a low-temperature chamber. The sample was oxidized by the air in the bottle, and we termed this process as natural oxidation. It is important to note that the bottle was covered with a cap but not sealed. The same initial sample was subjected to different oxidation times and labeled as SA1 (as-synthesis), SA2 (1–2 months) and SA3 (3–6 months). The color of the CD solution clearly changed from dark brown to the light brown during the oxidation process, as displayed in Figure 1. It is essential to clarify that these three samples are not prepared in separate batches (except the samples used for the measurements of nuclear magnetic resonance); rather, the same initial sample underwent different oxidation periods, which were utilized for the microstructure characterizations and optical measurements. It is worth noting that even for the CDs from different batches, their luminescent characteristics will undergo the aforementioned three stages during the long period of oxidation, and the changing process is also repeatable.

Figure 1. Photographs of (**a**) SA1, (**b**) SA2, and (**c**) SA3 under the irradiation of daylight.

2.2. Characterization and Optical Experiments

Transmission electron microscopy (TEM) images were obtained by utilizing a FEI Tecnai G2 F30 TEM (Thermo Fisher Scientific, Waltham, MA, USA) operating at a voltage of 300 kV. Fourier-transform infrared (FTIR) spectra were measured using a TENSOR II FTIR Spectrometer (Bruker, Ettlingen, Germany) with a spectral resolution of 4 cm^{-1} in the range of 4000 to 400 cm^{-1}. X-ray photoelectron spectroscopy (XPS) analysis was performed on an Escalab 250Xi (Thermo Electron, Waltham, MA, USA) using Al Kα (1486.6 eV) monochromatic radiation via the excitation source. Raman spectra ranging from 1000 to 1500 cm^{-1} were recorded via a LabRAM HR Evolution micro Raman system (HORIBA Scientific, Vénissieux, France) with a spectral resolution of 2 cm^{-1}, utilizing a diode-pumped solid-state laser with the 532 nm line as the excitation source. X-band electron paramagnetic resonance (EPR) spectra were acquired via a Bruker EMX-plus spectrometer (Bruker, Germany) operating at 9.4 GHz under the center field of 3250 G. The ^{1}H and ^{13}C nuclear magnetic resonance (NMR) spectra were collected using a Bruker AM X400 (Bruker, Germany) spectrometer operating at a frequency of 400 and 100 MHz, respectively.

The UV-vis absorbance spectra ranging from 220 to 600 nm were obtained using a Shimazu UV-3600 spectrophotometer (Shimadzu, Kyoto, Japan). The steady photoluminescence (PL) spectra were measured using an F-7000 (Hitachi, Tokyo, Japan) fluorescence spectrometer with a 450 W Xenon lamp as the excitation source. Both the excitation and emission slits were set to 2.5 nm, and the photomultiplier tube (PMT) was utilized to detect the emission light. Time-resolved photoluminescence (TRPL) spectra were recorded with an FLS980 fluorescence spectrophotometer (Edinburgh, UK). The equipped excitation lasers were an EP-LED-360 picosecond (ps) pulse LED (λ_{exc} = 360 nm and pulse duration ~950 ps) and an EPL-450 ps pulse diode laser (λ_{exc} = 405 nm and pulse duration ~75 ps), respectively. The absolute fluorescence QY was measured using an FLS980 fluorescence spectrophotometer with an integrating sphere.

3. Results

3.1. Characterizations of the CDs

Figure 2 shows the TEM images of SA1, SA2, and SA3 with different oxidation degrees. The high-resolution TEM (HR-TEM) images of the CDs display the well-resolved lattice fringes, and the corresponding interplane spacing for the three samples is about 0.21 nm (upper left insets), which is close to the value of the (100) plane of graphitic carbon [7,8]. The particle size distributions of the three samples range from 2 to 6 nm, and the CDs exhibit a quasi-spherical shape with an average diameter of ~3 nm. The CDs retain good solubility in water even after a long period of oxidation, as displayed in the Supplementary Materials (Figure S1). However, due to the relatively wide size distribution of the dots, it is challenging to discern the variations in their size and shape during the long oxidation time.

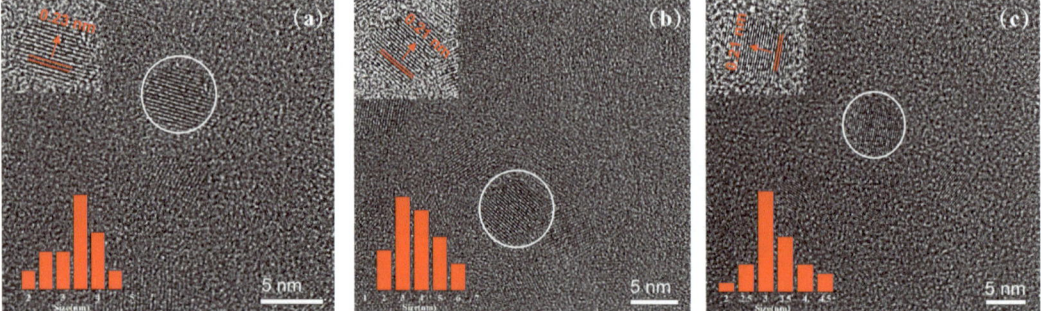

Figure 2. TEM images of the (**a**) SA1, (**b**) SA2, and (**c**) SA3 samples, and the upper and lower left insets are the HRTEM images (the circled CDs) and size distribution of the CDs, respectively.

The chemical bonding and surface configurations of the CDs can be derived from FTIR and Raman spectra measurements, as shown in Figure 3. In the FTIR spectra (Figure 3a), three main absorption peaks at approximately 1045, 1394, and 1579 cm^{-1} are observed, corresponding to the C–O–C [21], C–O [22], and C=C functional groups [21], respectively. In addition, a broad absorption band centered around 3286 cm^{-1} indicates the presence of the stretching vibrations of C–OH bonding [23]. Several small peaks at around 1243, 1448, 1700, and 2942 cm^{-1} arise from the vibrational absorption of C–O [18], C–H(CH$_2$) [24], C=O/COOH [25,26], and C–H (CH$_3$) [24] groups, respectively. The presence of the numerous hydrophilic groups endows the CDs with good solubility in water.

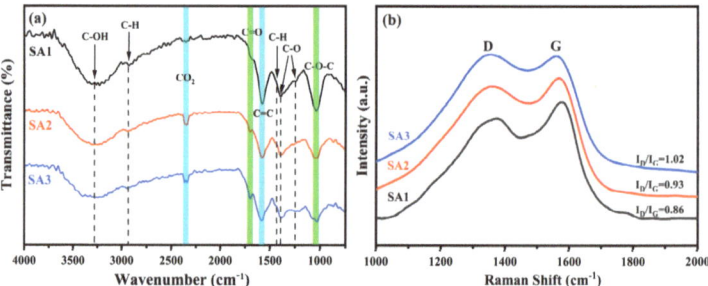

Figure 3. (**a**) FTIR and (**b**) Raman spectra of SA1 (black curve), SA2 (red curve), and SA3 (blue curve).

To get the FTIR signal, the liquid CDs were freeze-dried into powders. After a long period of oxidation, the C=O (1700 cm^{-1}) signal for SA3 becomes slightly more pronounced due to the weakened absorption of the C=C bonds. However, its absorption intensity has not changed significantly. Additionally, the absorption intensities of C–O–C and C–O decrease from the SA1 to the SA3 sample, which may arise from the different concentrations of CDs used when measured, which will be further discussed in the UV-absorption section. Additionally, a new absorption peak at approximately 2347 cm^{-1}, associated with CO$_2$, appears in the SA2 and SA3 samples. This may result from CO$_2$ adsorption or the high oxidation degree on the surface of the CDs [27].

Figure 3b shows the Raman spectra of SA1, SA2, and SA3. Two peaks are observed around 1373 cm^{-1} (D-band) and 1578 cm^{-1} (G-band). The G-band is associated with the graphitic sp^2 carbons, whereas the D-band is related to the sp^3 disorder carbons [15,27]. Notably, the intensity ratio I_D/I_G increases from 0.86 to 1.02 with longer oxidation time, indicating that more oxygen atoms incorporate into the CD structures and form oxygen-containing groups [15,27].

To elucidate the evolution of the elemental composition and chemical groups of CDs during different oxidation stages, XPS measurements were carried out. Figure 4 displays the C 1s and O 1s XPS spectra of the SA1, SA2, and SA3 samples, from which the percentage of C and O atoms could be calculated. The fractions of O (C) content in SA1, SA2, and SA3 are 30.95% (69.05%), 39.88% (60.12%), and 42.40% (57.60%), respectively. This demonstrates that the as-synthesized CDs (SA1) already contain a considerable number of oxygen atoms, and the oxygen content in the CDs increases gradually when the natural oxidation time is prolonged, indicating that many more O atoms participated in the formation of chemical groups at the surface of the CDs.

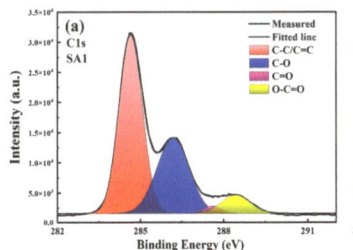
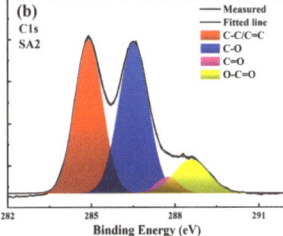
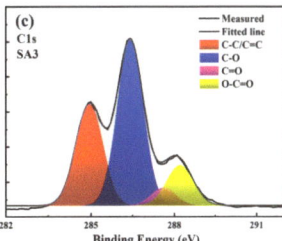

Figure 4. The C 1s high-resolution XPS spectra of (**a**) SA1, (**b**) SA2, and (**c**) SA3.

The high-resolution C 1s XPS spectra of the three samples can be deconvoluted into four Gaussian components: C=C/C–C (graphitic carbon) at ~284.8 eV [28], C–O at ~286.2 eV (epoxy) [15], C=O at ~287.6 eV(carbonyl) [29], and O–C=O at ~288.5 eV (carboxyl) [30,31]. The percentages of graphitic carbon and oxygenated carbon are listed in Table 1. As the oxidation time increases, there is a gradual decline in the proportion of graphitic carbon from 48.9% to 30.4%, while the content of oxygenated carbon (C–O, C=O and O–C=O) increases. The progressively increasing presence of the C=O groups is consistent with the FTIR analysis. However, no obvious change is observed in the O–C=O groups between the SA2 and SA3 samples. The high-resolution O 1s XPS spectra of the SA1 and SA2 samples consist of two fitting peaks at around 532.7 eV and 531.4 eV (Figure S2), which are attributed to the C=O and C–OH/C–O–C groups [31], respectively. Their ratios are listed in Table S1. Moreover, for the SA3 sample, in addition to the C=O and C–OH/C–O–C groups, a new band centered at 536 eV appears, which could be attributed to the absorption of H_2O or O_2 [32,33].

Table 1. The XPS data analyses of the C 1s spectra.

	C–C/C=C	C–O	C=O	O–C=O
SA1	48.9%	33.2%	4.8%	13.1%
SA2	37.2%	39.2%	7.3%	16.3%
SA3	30.4%	44.2%	9.3%	16.1%

The ^{1}H and ^{13}C NMR spectra of CDs in DMSO further confirm the presence of abundant by-products during the synthetic process, as shown in Figure S3. In the ^{1}H NMR spectra (Figure S3a), two main regions can be observed: one in the range of 1–2.5 ppm, corresponding to the sp^3 C–H protons, and the other between 3 and 6 ppm, originating from the protons attached to the hydroxyl, ether, and carbonyl groups [34]. In addition, the peak at 8.44 ppm could be attributed to the formate [35]. In the ^{13}C NMR spectra (Figure S3b), four main regions are identified: peaks between 20 and 80 ppm originating from aliphatic (sp^3) carbon atoms [34], signals from 80 to 85 ppm attributed to the carbons linked with ether groups [34], peaks in the range of 90–102 ppm arising from the anomeric carbons of α-pyranoses [36], and signals between 165 and 180 ppm ascribed to the carboxyl and carbonyl groups [34,37]. Additionally, the peak near 105 ppm is due to the anomeric carbons in β-pyranoses [36].

The analysis of the ^{1}H and ^{13}C NMR spectra reveals that in addition to the presence of hydroxyl, ether, and carbonyl groups, by-products such as formate and pyranoses are also detected. This suggests that the synthesized CD solution contains numerous organic compounds. A multitude of by-products are generated during the synthesis process of CDs, which causes their NMR signals to cluster together, as depicted in Figure S3. For the NMR measurements, the samples were dealt with spin-drying to substantially reduce the effects of water on the results. In addition, the concentration of small carbonaceous materials gradually decreases during the oxidation process. Consequently, even if the microstructure of the sample changes during the oxidation process, distinguishing these changes remains challenging due to the inability to control the concentration of CDs and small carbonaceous materials when measured.

3.2. Steady Optical Properties

Figure 5 displays the UV-Vis absorbance spectra of the three samples, the main absorption peak at 265 nm can be attributed to the π–π* transitions of C=C, while the shoulder band centered at approximately 354 nm stems from the n–π* transitions of C=O [20,31,38]. It is evident that the overall absorbance intensity significantly decreases with the increasing oxidation time.

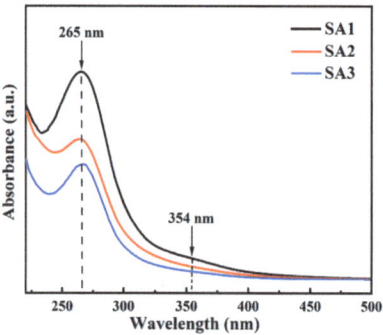

Figure 5. The UV-vis absorbance spectra of SA1 (black curve), SA2 (red curve), and SA3 (blue curve).

According to the Raman and XPS analyses, as the oxidation degree of the CDs increases, the content of the graphitic carbon decreases, accompanied by an increase in the fraction of oxygenated carbon. Although this could explain the weakening of π–π* transitions in the graphitic carbon (C=C), it is difficult to interpret the reduced absorption intensity associated with the C=O groups, which seemingly contradicts the increment of the carbonyl and carboxyl groups. During the synthesis of the CDs, we consider that a significant amount of small carbonaceous materials (<1 nm) were generated when the glucose underwent condensation, dehydration, polymerization, or carbonization processes [10,11]. Parts of these small carbonaceous materials (including aromatic rings, small carbon clusters, and CDs with smaller dot size), with oxygen-containing groups might be transformed into a CO_2 structure after long-time oxidation. This transformation could lead to a decrease in CD concentration, which may be the main reason for the decrease in the overall absorption intensity as well as the reduction in the amount of the C–O–C and C–O contents observed in the FTIR measurements. In addition, the increasing transparency of the CD solution can further support this hypothesis.

It is well established that the PL properties of the CDs can be significantly modified by changing the oxidation degree of the CDs' surface structure, leading to variation in the electronic structure or emission states. Figure 6 illustrates the PL spectra of SA1, SA2, and SA3 under different excitation wavelengths, it is interesting to observe that the evolution of the PL spectra of CDs during the oxidation process can be evidently divided into three stages. For the as-synthesized sample (SA1), the PL intensity of the green emission (~505 nm) is stronger than that of the blue luminescence (~435 nm), as depicted in Figures 6a and S4a. In this excitation region, the peaks of the two emission bands change very little, while their intensity gradually increases. The optimal excitation wavelengths for the two emission peaks are about 392 and 422 nm, respectively, which could be acquired in the PL excitation (PLE) spectra as depicted in Figure S5a.

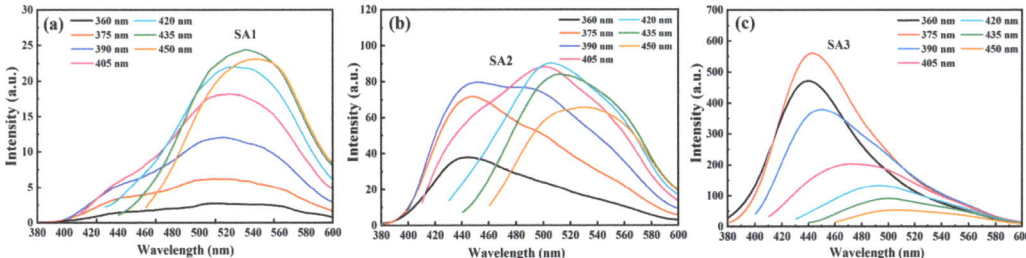

Figure 6. The PL spectra of (**a**) SA1, (**b**) SA2, and (**c**) SA3 under different excitation wavelengths.

Compared with the SA1 sample, the intensity of the blue emission band in the SA2 is stronger than that of the green luminescence under 360 nm excitation, as illustrated in Figure 6b. The PL intensity of both two emission bands increases within the excitation range between 360 and 390 nm, but the intensity of the green luminescence becomes comparable to that of the blue emission when excited by 390 nm. An obvious red-shift in the green luminescence is observed when the excitation wavelength is further increased.

The SA3 sample exhibits clear excitation-dependent behavior, the PL peak red-shifts with the increasing excitation wavelength. Notably, the blue emission dominates within the excitation range of 360–390 nm, reaching a maximum PL intensity under ~365 nm excitation. In addition, both blue and green emissions in SA3 exhibit nearly identical optimal excitation wavelengths (~365 nm), as depicted in Figure S5c. Conversely, the optimal excitation wavelengths for the blue and the green luminescence between the SA1 and SA2 samples differ significantly, as shown Figure S5.

More importantly, the PL intensity of the CDs remarkably enhances in the blue region after a long oxidation period, as displayed in Figure 7. However, after being increase several times initially, the intensity of the green emission remains relatively stable, even when the oxidation time is prolonged further. This demonstration underscores the distinct dominant emission states among the three samples due to their varied excitation and emission properties.

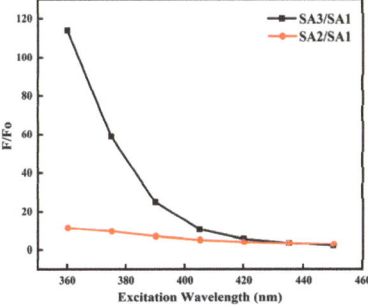

Figure 7. The ratio of the PL intensity of SA2/SA1 (red circle) and SA3/SA1 (black square) within the excitation range of 360–450 nm.

There are lots investigations about the PL mechanisms of dual-fluorescent CDs; several excitation and recombination models have been proposed to explain the luminescence processes, including the size effect (usually referred to as the sp^2 nanodomain), surface states, molecule states, the synergistic effect, etc. [10,11,39–42]. A large number of results demonstrate that the emission states induced by the oxygen-containing groups play a significant role in the optical properties of CDs [12–19,39–42]. However, these oxygenous functional groups have diverse structural compositions, including the hydroxyl (–OH), carbonyl (C=O), carboxyl (–COOH), and epoxy (C–O–C) groups. Due to the complex surface configurations, the origins of luminescence in CDs remain controversial, particularly for the blue and green emissions.

Wen et al. found that the fluorescence of CDs consisted of two distinct emission bands. One was associated with the intrinsic band located in the blue region, which was ascribed to the sp^2 sub domains. The other originated from the surface states introduced by abundant oxygen-containing functional groups, emitting longer-wavelength lights [43]. Wang's group suggested that the hydroxyl group decorated at the surface of graphene quantum dots (GQDs) played an important role in the blue fluorescence (430 nm), while the synergy effect of the hydroxyl group and the adjacent carbonyl group might cause the green emission (530 nm) [44]. Li and co-workers proposed that the blue emission centered at 440 nm was related to the -OH-related surface hybridized states, whereas the green or yellow-green was associated with the C–O–C (epoxy) groups [15].

In addition, numerous groups have hypothesized that oxygen-containing functional groups, acting as the non-radiative recombination centers, would suppress the intrinsic state emission associated with the sp^2 sub domains, leading to the low QY [12–14]. In contrast, several studies have suggested that the QY of CDs could be enhanced after an oxidation reaction [15–17,45]. For instance, Li's group reported that the PLQY of GQDs could be enhanced through oxidation treatment using H_2O_2 and UV light irradiation, with the emission color shifting from green to yellow-green, attributed to an increase in C–O–C (epoxy) groups [15]. Han et al. reported that the fluorescence QY of GQDs was increased through a post-oxidation method employing H_2O_2, resulting in a red-shift of the PL peak from 450 nm to 510 nm, ascribed to the increase in carbonyl and carboxyl groups [16]. Dong and co-workers found that the PL intensity of graphene oxide quantum dots (GOQDs) significantly decreased as the reduction degree increased, due to the decrease of the oxygen-containing groups, particularly the -COOH and -OH groups [45].

In our case, the broad PL band excited by the 375 nm could also be decomposed into a blue emission component and a green luminescence band, as depicted in Figure S4. Additionally, the QY of CDs increases from ~0.61% to ~4.26% after a long period of oxidation. The intensity of the blue emission is over one hundred times stronger than that of the as-synthesis sample, with the green luminescence also showing several-fold increases. According to the density functional theory (DFT) calculations conducted by Eda's group, the sp^2 cluster consisting of about five fused aromatic rings is responsible for emitting the blue light [46]. Sk et al. suggested that the emission wavelength could be tuned from ~400 to 572 nm by varying the dimeter of GQDs from 0.92 to 1.39 nm [47]. However, the TEM images reveal that after long-time oxidation, the particle sizes of the CDs still show a broad distribution ranging from 2 to 6 nm, with an average diameter of about 3 nm (>100 aromatic rings). Within these size ranges, the PL of CDs is predominantly in the red region. We speculate that the blue and green luminescence may originate from the CDs with smaller dot sizes or small carbonaceous materials.

To testify this hypothesis, we use the ultrafiltration treatment with a Millipore (3 kDa, cutoff) to filter the particles larger than 2 nm. The SA2 sample that underwent relatively short and long oxidation times were utilized, and their PL spectra after ultrafiltration treatment are displayed in Figure 8. Interestingly, the PL spectra of filtered SA2 samples at both oxidation stages exhibit similar excitation-dependent behaviors to those observed in the SA3 sample. Within the excitation range of 310–375 nm, there is a broad emission band spanning from 400 to 600 nm. Gaussian fitting allowed for the segmentation of this spectrum into two emission bands: a blue band centered at approximately 435 nm and a green band peaking around 490 nm, as depicted in Figure S4. The peak of the blue band remains almost unchanged under excitation wavelengths from 310 to 375 nm. Upon further increasing the excitation wavelength, the green luminescence peak gradually shifts towards longer wavelengths (red-shift). We also compare the PL spectra of the residual samples by the Millipore filter with those of the original solution (Figure S6), and find that the spectral features of the residue do not significantly differ from those of the original solution. Based on these ultrafiltration results, we consider that the blue emission primarily originates from small carbonaceous materials.

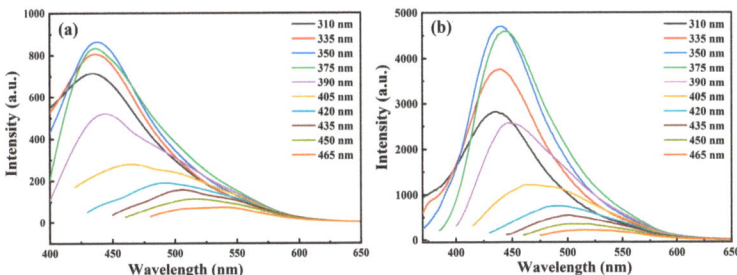

Figure 8. The PL spectra of the SA2 experienced relatively (**a**) short and (**b**) long oxidation times after ultrafiltration treatment when excited by different wavelengths. Both the samples are in the second oxidation stage.

Moreover, the change in the surface configurations and particle sizes of the CDs tailored through different oxidation methods (chemical or natural) may impart distinct optical properties to the CDs [14–17]. Xu's group used an HNO_3 reflexing method to oxidize the CDs and noted that the oxidation not only increased the oxygen content but also led to the formation of an oxygen-containing loose shell on the surface of the CDs [14]. Li et al. observed an increase in the amount of C-O-C bonds after they oxidized the graphene quantum dots (GQDs) with the hydrogen peroxide (H_2O_2) and UV light; notably, a high concentration of H_2O_2 could accelerate the oxidation process, potentially damaging or breaking down the conjugated structure of GQDs [15]. Additionally, Han and co-workers synthesized oxidized GQDs using a post-oxidation treatment with H_2O_2, which led to an increase in C=O and COOH groups while the proportion of C–O groups (epoxy and hydroxyl) decreased [16]. Srivastava's group altered the oxygen degree of the CDs via aerial oxidation, resulting in an increased density of hydroxyl groups on the CD surface and the transformation of C=N-H bonds into C=N bonds, with no significant changes to morphology [17].

In our study, we observed a gradual decline in the ratio of C–C/C=C, accompanied by an increase in the proportion of C–O and C=O groups, with the fraction of O–C=O groups reaching its maximum after an initial increase. We hypothesize that as the oxidation degree increases, the content of C–C/C=C components gradually diminishes, potentially leading to a slight reduction in the size of the small carbonaceous materials (<1 nm).

3.3. Time-Resolved PL (TRPL) Optical Properties

To further explore the luminescence process, measurements of the time-resolved PL (TRPL) spectra of CDs are carried out. Figure 9 displays the PL decay curves of three samples under ps LED light (λ_{exc} = 360 nm and pulse duration ~950 ps) excitation, the monitoring emission wavelengths are 430 and 510 nm, respectively. The PL decay curves could be well fitted by a bi-exponential function, typically used to evaluate the lifetime of CDs [48].

$$I(t) = A_1 \exp(-t/\tau_1) + A_2 \exp(-t/\tau_2) \qquad (1)$$

where A_1 and A_2 represent the amplitude of two decay components, respectively. The τ_1 and τ_2 are the lifetimes of two decay components, respectively. The average decay lifetime $\bar{\tau}$ can be calculated by the followed equation [48]:

$$\bar{\tau} = \frac{A_1 \tau_1^2 + A_2 \tau_2^2}{A_1 \tau_1 + A_2 \tau_2} \tag{2}$$

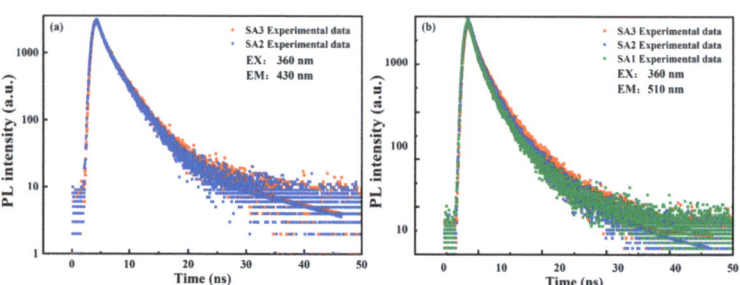

Figure 9. For time-resolved PL spectra of SA1, SA2, and SA3 with 360 nm as the excitation wavelength, the detected emission wavelengths are (**a**) 430 and (**b**) 510 nm, respectively.

The fitted and calculated values from Equations (1) and (2) are listed in Table 2. Analysis of the PL decay curves and data in Table 2 reveals a slight increase in the lifetime of CDs after a longer period of oxidation, with a decrease in the proportion of the shorter lifetime and an increase in the longer decay process. The PL dynamics of sample SA1 could not be detected due to the weak blue emission signal under the excitation of the ps LED pump source. For a comprehensive comparison of the recombination dynamics of the blue emission, the TRPL spectra of the three samples excited by a ps pulse diode laser (λ_{exc} = 405 nm and pulse duration ~75 ps) are also measured, as depicted in Figure S7. The detected emission wavelengths are 470 and 530 nm, respectively. The characteristics of the average lifetime and the proportions of the two decay processes are similar to those excited at 360 nm, except for the average lifetime of SA3 detected at 530 nm, which shows a difference (see Table S2).

Table 2. The fitting parameters and average lifetimes of the three samples under the excitation λ_{exc} = 360 nm.

Emission		430 nm			510 nm	
Sample	SA1	SA2	SA3	SA1	SA2	SA3
τ_1 (ns)	-	1.41	1.23	1.50	1.54	1.21
A_1 (%)	-	61.47	51.33	79.81	75.99	58.96
τ_2 (ns)	-	3.73	3.64	5.04	5.00	4.34
A_2 (%)	-	38.53	48.67	20.19	24.01	41.04
$\bar{\tau}$	-	2.90	3.01	3.12	3.29	3.44

Although several groups have also identified two distinct lifetimes in CDs, the origins of these decay processes are still a subject of debate [49–51]. Liu et al. attributed the fast component (~1.32 ns) to the recombination process involving the intrinsic states, and ascribed the slow lifetime (~7.89 ns) to the emissions from defect states [49]. However, Zhao and co-workers proposed that the faster component (1.29–1.73 ns) was associated with the radiative recombination of the eigenstates, while the slower lifetimes (2.51–7.16 ns) arising from the non-radiative process were related to the surface defects [50]. In addition, Byun's group suggested that the two lifetimes originated from the nonradiative recombination (1.17–3.15 ns) and radiative recombination processes (5.7–8.52 ns) of oxygen-induced defects, respectively [51].

3.4. Surface Defects and Luminescence Mechanisms of CDs

In order to deeply investigate the relationship between the microstructure and PL properties of CDs at different oxidation stages, the EPR spectra of the three samples are measured. The trend in the concentration of the unpaired electrons or free radicals is illustrated in Figure 10. For each measurement, about 5 mg of freeze-dried powders are used. It is evident that the intensity of the EPR signal decreases gradually from SA1 to SA3, indicating a reduction in the concentration of unpaired electrons or free radicals with an increasing oxidation time. The g-value, determined from the EPR spectrum, is approximately 2.005, typically associated with the oxygen-centered radicals [52,53]. Additionally, the EPR spectral linewidth is about 1 mT, suggesting that the free radicals predominantly arise from oxygen-containing groups connected to the sp^3 hybridization of carbon [54].

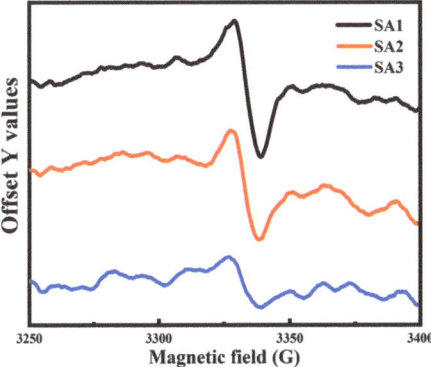

Figure 10. The EPR spectra of SA1 (black curve), SA2 (red curve), and SA3 (blue curve).

4. Discussion

To uncover the impact of oxygen on the luminescence mechanisms of CDs, several key issues need clarification. Firstly, after a long period of oxidation, the blue emission components of CDs become dominant under high-energy photon irradiation. Secondly, the blue emission displays an excitation-independent characteristic, whereas the green luminescence exhibits excitation-dependent behavior. Thirdly, the CD solution (SA2) obtained through ultrafiltration treatment shows PL characteristics similar to those of the sample with a higher oxidation degree.

First of all, according to the UV-vis absorbance and PLE spectra of CDs, the absorption peaks centered at 265 and 354 nm scarcely change. However, the optimal excitation wavelengths for the blue and green emissions undergo a blue shift after a long period of oxidation, eventually converging around 365 nm. This wavelength, where the maximum luminescence intensity is achieved, overlaps with the n–π* transitions of C=O. The close proximity of the optimal excitation wavelengths to the absorption peak may be a crucial factor in enhancing the PL intensity of the CDs. Additionally, both the blue and the green luminescence have two decay processes: fast and slow components. Notably, the PL dynamics of the detected blue and green fluorescence show minimal differences, suggesting that the luminescence centers possess similar recombination processes (lifetimes). Despite this, the QY of CDs with higher oxygen contents remains low compared to those with high-quality surface passivation. This observation leads to the hypothesis that the fast component is likely due to non-radiative recombination behaviors, while the slow component may result from the radiative recombination processes of photon-generated carriers.

Moreover, the blue emission was thought to arise from the small carbonaceous materials, as indicated by discussions of the PL properties of CDs after ultrafiltration treatment. The XPS results reveal that the proportions of C–O and C=O groups gradually increase when the oxidation time is prolonged, while the ratio of the O–C=O group reaches its maximum after an initial rise. It is noteworthy that the PL intensity of the blue emission exhibits a similar variation trend to the amounts of the C–O and C=O groups. Furthermore, the changing trend in the intensity of the green emission accords well with the variation in the O–C=O contents. Therefore, combining these analytical findings, we propose that the blue emission is associated with the C–O and C=O groups decorating the edges of small carbonaceous materials. Since the green luminescence has similar PL dynamics to the blue fluorescence, it is likely attributed to the O–C=O groups.

Righetto et al. synthesized three types of CDs (pCDs, oCDs, and mCDs) through solvothermal pyrolysis using ortho-, meta-, and para-phenylenediamine, respectively. The resulting CD solutions were carefully purified and subsequently characterized using fluorescence correlation spectroscopy and time-resolved electron paramagnetic resonance techniques. The findings revealed that the emission properties were predominantly dominated by free fluorescent molecular by-products [55]. By analyzing the ^{1}H and ^{13}C NMR spectra of CDs in DMSO, we find that the synthesized CD solution contains various organic compounds. These molecular products could impact the optical properties of CDs.

Secondly, the blue emission displays excitation-independent behavior, whereas the green luminescence is excitation-dependent. This suggests significant differences in the emission states and energy distributions within the bandgap of the carbon dots (CDs), induced by their respective oxygen-containing groups. Sun's group also observed similar phenomena and noted that CDs with different oxygen contents exhibit distinct excitation-dependent behaviors [56]. These observations were attributed to the competition between various transition processes arising from various oxygenous groups. It is speculated that the carboxyl as well as carbonyl functional groups are related to the green waveband, while the blue emission is possibly associated with the surface functional groups in the planes or at the edges of the CDs, such as the hydroxyl groups. In our case, the blue emission dominates the fluorescence when the carbonaceous materials have a smaller particle size or a higher oxidation degree, indicating that the small carbonaceous materials are more prone to oxidation due to their larger surface-to-volume ratio during the natural oxidation process. A substantial portion of carbon is transformed into the C–O and C=O groups, increasing the density of relevant emission states and significantly enhancing the blue emission intensity. The increase in the proportion of oxygen-related states at high energy levels may cause the blue shift in the optimal excitation wavelengths. Furthermore, the relatively uniform distribution of the C–O and C=O groups on the surface of small carbonaceous materials may contribute to the excitation-independent characteristics of the blue emission. Conversely, the intensity of the green emission is generally stronger in CDs with larger particle sizes or lower oxidation degrees. The non-uniform coverage of O–C=O groups might result in a wide distribution of the related energy levels, leading to excitation-dependent behavior.

Additionally, both concentration and aggregation are important factors affecting the optical properties of CDs. To illustrate this, the CD solutions are diluted 1-, 2-, 3-, 4-, and 5-fold with water, and the corresponding PL spectra are depicted in Figure S8. The sample exhibits a PL peak centered at approximately 440 nm under 360 nm of excitation, which is almost unchanged regardless of the dilution fold. The PL intensity reaches its maximum when the solution is diluted 3-fold, then decreases as the dilution fold increases further. Since the blue emission primarily stems from the oxygen-containing functional groups (C–O and C=O) on the surface of the small carbonaceous materials, the significant increase in PL intensity upon dilution highlights the presence of the aggregation within these materials. A broad PL band peaked at around 500 nm is obtained when the sample is excited at 420 nm. The PL peak experiences a slight blue-shift when the dilution fold is two, but remains almost unchanged with further dilution. The PL intensity decreases gradually as the dilution fold increases, yet this decrease is not directly proportional to the dilution fold. We attribute the green luminescence primarily to the oxygenous groups (O–C=O) decorated at the surface of the CDs with larger particle sizes or lower oxidation degrees.

Unlike the small carbonaceous materials, the aggregation phenomenon is less pronounced in CDs with larger particle sizes. Furthermore, the concentration of small carbonaceous materials decreases with the increase in the oxidation degree, which also contributes to a rise in PL intensity.

Finally, particle size plays a crucial role in influencing the PL properties of CDs. With the increasing oxidation degree, the surface of the CDs undergoes further oxidation, accompanying a reduction in CD size. This change can alter the coupling degree between the π-electron system and the oxygen-related surface states, leading to the blue shift in the PL peak. As illustrated in Figure 6, no blue shift is observed in the blue emission, likely due to the relatively broad size distribution of the CDs. It remains challenging to identify the effects of particle size variation on the PL peak shift. We implemented a straightforward ultrafiltration treatment using a Millipore (3 kDa, cutoff) to filter the particles larger than 2 nm. The PL spectra after ultrafiltration treatment are shown in Figure 8. When excited at 360 nm, the blue emission dominated the PL spectra, the spectral characteristics of the filtered sample differ markedly from those of the unfiltered sample. Nevertheless, the PL peak of the blue band remains nearly unchanged, indicating that even small carbonaceous materials possess a certain degree of particle size distribution.

For the residual samples in the ultrafiltration tube, their spectral characteristics remained identical to those of the original CD solution. One possible explanation for this is that the small carbonaceous materials are not entirely filtered out, with some still physically adhering to the surfaces of the CDs with larger dot sizes [39]. Another possibility is that the sp^{3-}- and sp^{2-}-hybridized carbons on the CD surface, decorated with the oxygen functional groups, may also emit blue and green light when excited.

Moreover, there still exists competition among different emission states and non-radiative traps, as illustrated in the schematic model of the excitation and recombination process in Figure 11. Initially, the photo-excited electrons are generated by transitioning from the n-state orbitals (HUMO) to the π* orbitals (LUMO) of C=O by absorbing the high-energy photons. These electrons quickly relax into oxygen-related emission states or get trapped by the non-radiative defects. The green and the blue luminescence stems from the radiative recombination of the electron-hole pairs at corresponding oxygen-related energy levels. The EPR spectra confirm that the concentration of oxygen-centered radicals would decrease after a long period of oxidation, potentially reducing the non-radiative traps and enhancing the QY of CDs. However, only a slight increase in the average lifetime is observed in samples with higher oxygen contents. We speculate that the increase in oxygenous groups may enhance the density of oxygen-related emission states and improve the relaxation rate between energy levels. This could counterbalance the effect of reduced non-radiative traps on the recombination process, resulting in only a small change in the lifetime of the CDs.

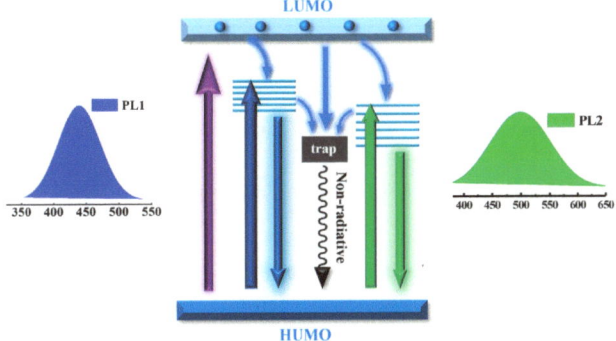

Figure 11. A schematic diagram of the PL mechanisms of the oxidized-CDs.

5. Conclusions

The dual-fluorescent (blue and green emission) CDs are prepared using a simple ultrasound method. The surface configuration of the CDs could be softly altered through natural oxidation. The QY of the CDs increases significantly from approximately 0.61% to 4.26% after a long period of oxidation. Based on microstructural characterizations, optical measurements, and ultrafiltration experiments, we consider that the C–O and C=O groups are prone to form at the surface of small carbonaceous materials during the natural oxidation process, leading to the enhancement of the blue emission. Conversely, the green emission, associated with O–C=O groups, is stronger in CDs with larger particle sizes or lower oxidation degrees. The emission states induced by their corresponding oxygenous groups display distinct distribution and energy levels within the bandgap of small carbonaceous materials or CDs, leading to excitation-independent blue emission and excitation-dependent green luminescence. Furthermore, the increase in emission states caused by corresponding oxygen-containing groups as well as the reduction of the amount of oxygen-centered radicals contribute to the enhanced QYs of the CDs. This work provides an in-depth understanding of the role of oxygen in the optical properties of CDs.

Supplementary Materials: The following supporting information can be downloaded at: https://www.mdpi.com/article/10.3390/nano14110970/s1, Figure S1: TEM images of (a) SA1, (b) SA2, and (c) SA3 (50 nm scale bar); Figure S2: O1s high-resolution XPS spectra of (a) SA1, (b) SA2, and (c) SA3; Figure S3: (a) ^{1}H NMR and (b) ^{13}C NMR spectra of SA1 (black line), SA2 (red line), and SA3 (blue line); Figure S4: The Gaussian fitting results of PL spectra excited by 375 nm for (a) SA1, (b) SA2, and (c) SA3; Figure S5: PLE spectra of (a) SA1, (b) SA2, and (c) SA3. The emission wavelengths are 435 (black line) and 505 nm (red line), respectively; Figure S6: The PL spectra of the residual samples for SA2 experienced relatively (a) short and (b) long oxidation times; The PL spectra of the SA2 experienced (c) short and (d) relative long oxidation times before ultrafiltration treatment; Figure S7: Time-resolved PL spectra of SA1, SA2, and SA3 with the 405 nm as the excitation wavelength, the emission wavelengths are (a) 470 and (b) 530 nm, respectively; Figure S8: The PL spectra of SA2 at various dilution levels (1, 2, 3, 4 and 5-fold) under (a) 360 nm and (b) 420 nm of excitation; (c) The PL intensity of SA2 under 360 nm (black square) and 420 nm (red circle) of excitation as a function of the dilution fold; Table S1: XPS data analyses of the O 1s spectra; Table S2: The fitting parameters and average lifetimes of the three samples under the excitation λ_{exc} = 405 nm.

Author Contributions: Conceptualization, P.Z., Y.Z. and L.R.; methodology, P.Z., Y.Z. and S.L.; validation, M.F. and Q.Z.; formal analysis, R.Q. and Z.Q.; investigation, P.Z., S.L. and Q.Z.; resources, J.Z.; writing—original draft preparation, P.Z.; writing—review and editing, L.R. and L.J.; visualization, Z.Q.; supervision, J.Z.; funding acquisition, L.J. All authors have read and agreed to the published version of the manuscript.

Funding: This work was funded by the National Natural Science Foundation of China, grant numbers 12004348 and 62073299, the Joint Fund of the Henan Province Science and Technology R&D Program, grant number 225200810071, the Central Plains Science and Technology Innovation Leaders Project, grant number 224200510026, the Key scientific research project of Henan University, grant number 24A535001, and the Henan Province Science and Technology Plan Project, grant number 232102220008.

Data Availability Statement: The data presented in this work are available upon request via email to the corresponding author.

Conflicts of Interest: The authors declare no conflicts of interest.

References

1. Zhang, H.; Wang, G.; Zhang, Z.; Lei, J.H.; Liu, T.M.; Xing, G.; Deng, C.; Tang, Z.; Qu, S. One step synthesis of efficient red emissive carbon dots and their bovine serum albumin composites with enhanced multi-photon fluorescence for in vivo bioimaging. *Light Sci. Appl.* **2022**, *11*, 113. [CrossRef] [PubMed]
2. Wang, B.; Cai, H.; Waterhouse, G.I.N.; Qu, X.; Yang, B.; Lu, S. Carbon dots in bioimaging, biosensing and therapeutics: A comprehensive review. *Small Sci.* **2022**, *2*, 2200012. [CrossRef]

3. Zhang, Y.; Sun, M.; Lu, Y.; Peng, M.; Du, E.; Xu, X. Nitrogen-doped carbon dots encapsulated a polyoxomolybdate-based coordination polymer as a sensitive platform for trace tetracycline determination in water. *Nanomaterials* **2023**, *13*, 2676. [CrossRef] [PubMed]
4. Shi, Y.; Su, W.; Yuan, F.; Yuan, T.; Song, X.; Han, Y.; Wei, S.; Zhang, Y.; Li, Y.; Li, X.; et al. Carbon dots for electroluminescent light-emitting diodes: Recent progress and future prospects. *Adv. Mater.* **2023**, *35*, 2210699. [CrossRef] [PubMed]
5. Zhang, Q.; Wang, R.; Feng, B.; Zhong, X.; Ostrikov, K.K. Photoluminescence mechanism of carbon dots: Triggering high-color-purity red fluorescence emission through edge amino protonation. *Nat. Commun.* **2021**, *12*, 6856. [CrossRef] [PubMed]
6. Li, J.; Chen, J.; Zhao, X.; Vomiero, A.; Gong, X. High-loading of organosilane-grafted carbon dots in high-performance luminescent solar concentrators with ultrahigh transparency. *Nano Energy* **2023**, *115*, 108674. [CrossRef]
7. Zhao, H.; Liu, G.; You, S.; Camargo, F.V.A.; Zavelani-Rossi, M.; Wang, X.; Sun, C.; Liu, B.; Zhang, Y.; Han, G.; et al. Gram-scale synthesis of carbon quantum dots with a large Stokes shift for the fabrication of eco-friendly and high-efficiency luminescent solar concentrators. *Energy Environ. Sci.* **2021**, *14*, 396–406. [CrossRef]
8. Zhang, Y.; Wang, L.; Hu, Y.; Sui, L.; Cheng, L.; Lu, S. Centralized excited states and fast radiation transitions reduce laser threshold in carbon dots. *Small* **2023**, *19*, 2207983. [CrossRef]
9. Zhang, Y.; Song, H.; Wang, L.; Yu, J.; Wang, B.; Hu, Y.; Zang, S.Q.; Yang, B.; Lu, S. Solid-state red laser with a single longitudinal mode from carbon dots. *Angew. Chem. Int. Ed.* **2021**, *60*, 25514–25521. [CrossRef]
10. Ai, L.; Yang, Y.; Wang, B.; Chang, J.; Tang, Z.; Yang, B.; Lu, S. Insights into photoluminescence mechanisms of carbon dots: Advances and perspectives. *Sci. Bull.* **2021**, *66*, 839–856. [CrossRef]
11. Xue, S.; Li, P.; Sun, L.; An, L.; Qu, D.; Wang, X.; Sun, Z. The formation process and mechanism of carbon dots prepared from aromatic compounds as precursors: A review. *Small* **2023**, *19*, 2206180. [CrossRef] [PubMed]
12. Zheng, H.; Wang, Q.; Long, Y.; Zhang, H.; Huang, X.; Zhu, R. Enhancing the luminescence of carbon dots with a reduction pathway. *Chem. Commun.* **2011**, *47*, 10650–10652. [CrossRef] [PubMed]
13. Mei, Q.; Zhang, K.; Guan, G.; Liu, B.; Wang, S.; Zhang, Z. Highly efficient photoluminescent graphene oxide with tunable surface properties. *Chem. Commun.* **2010**, *46*, 7319–7321. [CrossRef] [PubMed]
14. Xu, Y.; Wu, M.; Feng, X.Z.; Yin, X.B.; He, X.W.; Zhang, Y.K. Reduced carbon dots versus oxidized carbon dots: Photo- and electrochemiluminescence investigations for selected applications. *Chem.-A Eur. J.* **2013**, *19*, 6282–6288. [CrossRef] [PubMed]
15. Li, Y.; Liu, X.; Wang, J.; Liu, H.; Li, S.; Hou, Y.; Wan, W.; Xue, W.; Ma, N.; Zhang, J.Z. Chemical nature of redox-controlled photoluminescence of graphene quantum dots by post-synthesis treatment. *J. Phys. Chem. C* **2016**, *120*, 26004–26011. [CrossRef]
16. Han, T.; Zhou, X.; Wu, X. Enhancing the fluorescence of graphene quantum dots with a oxidation way. *Adv. Mat. Res.* **2014**, *887–888*, 156–160. [CrossRef]
17. Srivastava, I.; Misra, S.K.; Tripathi, I.; Schwartz-Duval, A.; Pan, D. In situ time-dependent and progressive oxidation of reduced state functionalities at the nanoscale of carbon nanoparticles for polarity-driven multiscale near-infrared imaging. *Adv. Biosyst.* **2018**, *2*, 1800009. [CrossRef]
18. Liu, C.; Bao, L.; Yang, M.; Zhang, S.; Zhou, M.; Tang, B.; Wang, B.; Liu, Y.; Zhang, Z.; Zhang, B.; et al. Surface sensitive photoluminescence of carbon nanodots: Coupling between the carbonyl group and π-electron system. *J. Phys. Chem. Lett.* **2019**, *10*, 3621–3629. [CrossRef]
19. Ding, H.; Yu, S.B.; Wei, J.S.; Xiong, H.M. Full-color light-emitting carbon dots with a surface-state-controlled luminescence mechanism. *ACS Nano* **2016**, *10*, 484–491. [CrossRef]
20. Li, H.T.; He, X.D.; Liu, Y.; Huang, H.; Lian, S.Y.; Lee, S.T.; Kang, Z.H. One-step ultrasonic synthesis of water-soluble carbon nanoparticles with excellent photoluminescent properties. *Carbon* **2011**, *49*, 605–609. [CrossRef]
21. Li, M.; Cushing, S.K.; Zhou, X.; Guo, S.; Wu, N. Fingerprinting photoluminescence of functional groups in graphene oxide. *J. Mater. Chem. A* **2012**, *22*, 23374–23379. [CrossRef]
22. Li, L.L.; Ji, J.; Fei, R.; Wang, C.Z.; Lu, Q.; Zhang, J.R.; Jiang, L.P.; Zhu, J.J. A facile microwave avenue to electro chemiluminescent two-color graphene quantum dots. *Adv. Funct. Mater.* **2012**, *22*, 2971–2979. [CrossRef]
23. Zhu, S.; Zhang, J.; Tang, S.; Qiao, C.; Wang, L.; Wang, H.; Liu, X.; Li, B.; Li, Y.; Yu, W.; et al. Surface chemistry routes to modulate the photoluminescence of graphene quantum dots: From fluorescence mechanism to up-conversion bioimaging applications. *Adv. Funct. Mater.* **2012**, *22*, 4732–4740. [CrossRef]
24. Nie, H.; Li, M.; Li, Q.; Liang, S.; Tan, Y.; Sheng, L.; Shi, W.; Zhang, S.X.-A. Carbon dots with continuously tunable full-color emission and their application in ratio metric pH sensing. *Chem. Mater.* **2014**, *26*, 3104–3112. [CrossRef]
25. Jang, M.-H.; Ha, H.D.; Lee, E.-S.; Liu, F.; Kim, Y.-H.; Seo, T.S.; Cho, Y.-H. Is the chain of oxidation and reduction process reversible in luminescent graphene quantum dots? *Small* **2015**, *11*, 3773–3781. [CrossRef] [PubMed]
26. Stan, C.S.; Elouakassi, N.; Albu, C.; Ania, C.O.; Coroaba, A.; Ursu, L.E.; Popa, M.; Kaddami, H.; Almaggoussi, A. Photoluminescence of argan-waste-derived carbon nanodots embedded in polymer matrices. *Nanomaterials* **2024**, *14*, 83. [CrossRef] [PubMed]
27. Fang, Y.; Zhao, Z.; Weng, Z.; Zhu, M.; Lei, W.; Zhu, Z.; Shafie, S.B.; Mohtar, M.N. Variation in the optical properties of carbon dots fabricated by a green and facile strategy for solar-blind UV detection. *J. Phys. Chem. C* **2022**, *126*, 5711–5721. [CrossRef]
28. Zhu, P.; Tan, K.; Chen, Q.; Xiong, J.; Gao, L. Origins of efficient multiemission luminescence in carbon dots. *Chem. Mater.* **2019**, *31*, 4732–4742. [CrossRef]

29. Bao, L.; Liu, C.; Zhang, Z.L.; Pang, D.W. Photoluminescence-tunable carbon nanodots: Surface-state energy-gap tuning. *Adv. Mater.* **2015**, *27*, 1663–1667. [CrossRef]
30. Sun, S.; Zhang, L.; Jiang, K.; Wu, A.; Lin, H. Toward high-efficient red emissive carbon dots: Facile preparation unique properties and applications as multifunctional theragnostic agents. *Chem. Mater.* **2016**, *28*, 8659–8668. [CrossRef]
31. Wei, S.; Yin, X.; Li, H.; Du, X.; Zhang, L.; Yang, Q.; Yang, R. Multi-color fluorescent carbon dots: Graphitized sp^2 conjugated domains and surface state energy level co-modulate band gap rather than size effects. *Chem. Eur. J.* **2020**, *26*, 8129–8136. [CrossRef] [PubMed]
32. Xu, X.; Zhang, K.; Zhao, L.; Li, C.; Bu, W.; Shen, Y.; Gu, Z.; Chang, B.; Zheng, C.; Lin, C.; et al. Aspirin-based carbon dots, a good biocompatibility of material applied for bioimaging and anti-inflammation. *ACS Appl. Mater. Interfaces* **2016**, *8*, 32706–32716. [CrossRef] [PubMed]
33. Danial, W.H.; Abdullah, M.; Abu Bakar, M.A.; Yunos, M.S.; Ibrahim, A.R.; Iqbal, A.; Adnan, N.N. The valorization of grass waste for the green synthesis of graphene quantum dots for nonlinear optical applications. *Opt. Mater.* **2022**, *132*, 112853. [CrossRef]
34. De, B.; Karak, N. A green and facile approach for the synthesis of water soluble fluorescent carbon dots from banana juice. *RSC Adv.* **2013**, *3*, 8286–8290. [CrossRef]
35. Whitty, S.D.; Waggoner, D.C.; Cory, R.M.; Kapla, L.A.; Hatcher, P.G. Direct noninvasive ^{1}H NMR analysis of stream water DOM: Insights into the effects of lyophilization compared with whole water. *Mag. Reson. Chem.* **2021**, *59*, 540–553. [CrossRef]
36. Wan, C.; Jiang, H.; Tang, M.T.; Zhou, S.; Zhou, T. Purification, physicochemical properties and antioxidant activity of polysaccharides from Sargassum fusiforme by hydrogen peroxide/ascorbic acid-assisted extraction. *Int. J. Biol. Macromol.* **2022**, *223*, 490–499. [CrossRef]
37. Song, Z.; Quan, F.; Xu, Y.; Liu, M.; Cui, L.; Liu, J. Multifunctional N, S co-doped carbon quantum dots with pH- and thermo-dependent switchable fluorescent properties and highly selective detection of glutathione. *Carbon* **2016**, *104*, 169–178. [CrossRef]
38. Ren, J.; Weber, F.; Weigert, F.; Wang, Y.; Choudhury, S.; Xiao, J.; Lauermann, I.; Resch-Genger, U.; Bande, A.; Petit, T. Influence of surface chemistry on optical chemical and electronic properties of blue luminescent carbon dots. *Nanoscale* **2019**, *11*, 2056–2064. [CrossRef]
39. Ding, H.; Li, X.H.; Chen, X.B.; Wei, J.S.; Li, X.B.; Xiong, H.M. Surface states of carbon dots and their influences on luminescence. *J. Appl. Phys.* **2020**, *127*, 231101. [CrossRef]
40. Gan, Z.; Xu, H.; Hao, Y. Mechanism for excitation-dependent photoluminescence from graphene quantum dots and other graphene oxide derivates: Consensus, debates and challenges. *Nanoscale* **2016**, *8*, 7794–7807. [CrossRef]
41. Gan, Z.; Xiong, S.; Wu, X.; Xu, T.; Zhu, X.; Gan, X.; Guo, J.; Shen, J.; Sun, L.; Chu, P.K. Mechanism of photoluminescence from chemically derived graphene oxide: Role of chemical reduction. *Adv. Opt. Mater.* **2013**, *1*, 926–932. [CrossRef]
42. Liu, Y.; Liang, F.; Sun, J.; Sun, R.; Liu, C.; Deng, C.; Seidi, F. Synthesis Strategies, Optical mechanisms, and applications of dual-emissive carbon dots. *Nanomaterials* **2023**, *13*, 2869. [CrossRef] [PubMed]
43. Wen, X.; Yu, P.; Toh, Y.R.; Hao, X.; Tang, J. Intrinsic and extrinsic fluorescence in carbon nanodots: Ultrafast time-resolved fluorescence and carrier dynamics. *Adv. Opt. Mater.* **2013**, *1*, 173–178. [CrossRef]
44. Wang, L.; Zhu, S.; Wang, H.; Wang, Y.; Hao, Y.; Zhang, J.; Sun, H. Unraveling bright molecule-like state and dark intrinsic state in green-fluorescence graphene quantum dots via ultrafast spectroscopy. *Adv. Opt. Mater.* **2013**, *1*, 264–271. [CrossRef]
45. Dong, Y.; Zhang, S.; Shi, L.; Chen, Y.; Ma, J.; Guo, S.; Chen, X.; Song, H. The photoluminescence of step-wise reduced graphene oxide quantum dots. *Mater. Chem. Phys.* **2018**, *203*, 125–132. [CrossRef]
46. Eda, G.; Lin, Y.Y.; Mattevi, C.; Yamaguchi, H.; Chen, H.A.; Chen, I.S.; Chen, C.W.; Chhowalla, M. Blue photoluminescence from chemically derived graphene oxide. *Adv. Mater.* **2010**, *22*, 505–509. [CrossRef] [PubMed]
47. Sk, M.A.; Ananthanarayanan, A.; Huang, L.; Lim, K.H.; Chen, P. Revealing the tunable photoluminescence properties of graphene quantum dots. *J. Mater. Chem. C* **2014**, *2*, 6954–6960. [CrossRef]
48. Hu, X.; An, X.; Li, L. Easy synthesis of highly fluorescent carbon dots from albumin and their photoluminescent mechanism and biological imaging applications. *Mater. Sci. Eng. C* **2016**, *58*, 730–736. [CrossRef] [PubMed]
49. Liu, F.; Jang, M.H.; Ha, H.D.; Kim, J.H.; Cho, Y.H.; Seo, T.S. Facile synthetic method for pristine graphene quantum dots and graphene oxide quantum dots: Origin of blue and green luminescence. *Adv. Mater.* **2013**, *25*, 3657–3662. [CrossRef]
50. Zhao, Y.; Yu, L.; Deng, Y.; Peng, K.; Yu, Y.; Zeng, X. A multi-color carbon quantum dots based on the coordinated effect of quantum size and surface defects with green synthesis. *Ceram. Int.* **2023**, *49*, 16647–16651. [CrossRef]
51. Byun, Y.; Jung, C.W.; Kim, J.H.; Kwon, W. Thermal control of oxygen-induced emission states in carbon dots for indoor lighting applications. *Dyes Pigm.* **2023**, *208*, 110895. [CrossRef]
52. Marciano, O.; Gonen, S.; Levy, N.; Teblum, E.; Yemini, R.; Nessim, G.D.; Ruthstein, S.; Elbaz, L. Modulation of oxygen content in graphene surfaces using temperature-programmed reductive annealing: Electron paramagnetic resonance and electrochemical study. *Langmuir* **2016**, *32*, 11672–11680. [CrossRef] [PubMed]
53. Rojas-Andrade, M.D.; Nguyen, T.A.; Mistler, W.P.; Armas, J.; Lu, J.E.; Roseman, G.; Hollingsworth, W.R.; Nichols, F.; Millhauser, G.L.; Ayzner, A.; et al. Antimicrobial activity of graphene oxide quantum dots: Impacts of chemical reduction. *Nanoscale Adv.* **2020**, *2*, 1074–1083. [CrossRef] [PubMed]
54. Genc, R.; Alas, M.O.; Harputlu, E.; Repp, S.; Kremer, N.; Castellano, M.; Colak, S.G.; Ocakoglu, K.; Erdem, E. High-capacitance hybrid supercapacitor based on multi-colored fluorescent carbon-dots. *Sci. Rep.* **2017**, *7*, 11222. [CrossRef] [PubMed]

55. Righetto, M.; Carraro, F.; Privitera, A.; Marafon, G.; Moretto, A.; Ferrante, C. The elusive nature of carbon nanodot fluorescence: An unconventional perspective. *J. Phys. Chem. C* **2020**, *124*, 22314–22320. [CrossRef]
56. Sun, Z.; Li, X.; Wu, Y.; Wei, C.; Zeng, H. Origin of green luminescence in carbon quantum dots: Specific emission bands originate from oxidized carbon groups. *New J. Chem.* **2018**, *42*, 4603–4611. [CrossRef]

Disclaimer/Publisher's Note: The statements, opinions and data contained in all publications are solely those of the individual author(s) and contributor(s) and not of MDPI and/or the editor(s). MDPI and/or the editor(s) disclaim responsibility for any injury to people or property resulting from any ideas, methods, instructions or products referred to in the content.

Article

Hydrophobic and Luminescent Polydimethylsiloxane PDMS-Y_2O_3:Eu^{3+} Coating for Power Enhancement and UV Protection of Si Solar Cells

Darya Goponenko [1,†], Kamila Zhumanova [1,†], Sabina Shamarova [1], Zhuldyz Yelzhanova [2], Annie Ng [2] and Timur Sh. Atabaev [1,*]

1. Department of Chemistry, School of Sciences and Humanities, Nazarbayev University, Astana 010000, Kazakhstan; darya.goponenko@nu.edu.kz (D.G.); kamila.zhumanova@nu.edu.kz (K.Z.); sabina.shamarova@nu.edu.kz (S.S.)
2. Department of Electrical and Computer Engineering, School of Engineering and Digital Sciences, Nazarbayev University, Astana 010000, Kazakhstan; zhuldyz.yelzhanova@nu.edu.kz (Z.Y.); annie.ng@nu.edu.kz (A.N.)
* Correspondence: timur.atabaev@nu.edu.kz; Tel.: +7-7172-70-60-26
† These authors contributed equally to this work.

Abstract: Solar cells have been developed as a highly efficient source of alternative energy, collecting photons from sunlight and turning them into electricity. On the other hand, ultraviolet (UV) radiation has a substantial impact on solar cells by damaging their active layers and, as a result, lowering their efficiency. Potential solutions include the blocking of UV light (which can reduce the power output of solar cells) or converting UV photons into visible light using down-conversion optical materials. In this work, we propose a novel hydrophobic coating based on a polydimethylsiloxane (PDMS) layer with embedded red emitting Y_2O_3:Eu^{3+} (quantum yield = 78.3%) particles for UV radiation screening and conversion purposes. The favorable features of the PDMS-Y_2O_3:Eu^{3+} coating were examined using commercially available polycrystalline silicon solar cells, resulting in a notable increase in the power conversion efficiency (PCE) by ~9.23%. The chemical and UV stability of the developed coatings were assessed by exposing them to various chemical conditions and UV irradiation. It was found that the developed coating can endure tough environmental conditions, making it potentially useful as a UV-protective, water-repellent, and efficiency-enhancing coating for solar cells.

Keywords: PDMS; Y_2O_3:Eu^{3+}; UV protection; hydrophobic coating; solar cells

Citation: Goponenko, D.; Zhumanova, K.; Shamarova, S.; Yelzhanova, Z.; Ng, A.; Atabaev, T.S. Hydrophobic and Luminescent Polydimethylsiloxane PDMS-Y_2O_3:Eu^{3+} Coating for Power Enhancement and UV Protection of Si Solar Cells. *Nanomaterials* **2024**, *14*, 674. https://doi.org/10.3390/nano14080674

Academic Editor: Zhixing Gan

Received: 8 March 2024
Revised: 8 April 2024
Accepted: 10 April 2024
Published: 12 April 2024

Copyright: © 2024 by the authors. Licensee MDPI, Basel, Switzerland. This article is an open access article distributed under the terms and conditions of the Creative Commons Attribution (CC BY) license (https://creativecommons.org/licenses/by/4.0/).

1. Introduction

Solar cells are playing an important role in global efforts to minimize reliance on conventional energy sources. These technologies not only help to reduce greenhouse gas emissions, but they also provide the potential to boost energy availability in rural or hard-to-reach regions. Despite well-established manufacturing technology, solar panels still face significant challenges that limit their efficiency, stability, and durability. In particular, spectral mismatches between solar cells' absorbance and solar radiation cause thermalization effects and loss of high energy (UV) and low energy (IR) photons [1,2]. Typically, the maximum conversion efficiency for crystalline silicon solar cells under the AM 1.5 solar spectrum is limited to around 29% [3]. Hence, anti-reflective coatings with various geometry and structure are commonly employed to surpass the Shockley–Queisser limit for single-junction devices [4,5]. On the other hand, another important factor to consider is the exposure of solar cells to UV radiation. Typically, UV radiation is not efficiently absorbed by silicon solar cells and contributes to thermalization and structural degradation processes, resulting in a quick decline in the performance of solar panels over time [6–8].

Thin film coatings composed of large-bandgap materials like TiO_2 and ZnO are commonly suggested to filter UV radiation [9,10]; however, this will result in the cutting-off of

UV photons from solar light. To address this issue, down-conversion (DC) optical materials capable of converting UV photons into several visible or infrared photons (quantum cutting effects) can be used [11–15]. Theoretical calculations revealed that a DC layer applied on the front surface of solar cells with Eg = 1.1 eV can boost efficiency by up to 38.6% as compared to 30.9% for cells without a coating [16]. However, under real-world conditions, the efficiency enhancements hardly surpass ~2–3%, which can be attributed to various light photon losses. For example, red-emitting $Sr_4Al_{14}O_{25}:Mn^{4+}$,Mg phosphor introduced in the polymethylmethacrylate (PMMA) layer increased the power conversion efficiency (PCE) of perovskite solar cells by ~1.4% [17]. Spin-coated CdSe/CdS quantum dots on the top of c-Si solar cells improved the PCE from 12% to 13.5% [18]. The PCE of commercial c-Si solar cells can also be improved by the liquid-phase deposition of cerium and ytterbium codoped $CsPbCl_{1.5}Br_{1.5}$ perovskite DC material [19]. Typically, the average PCE of c-Si solar cells was raised from 18.1% to 21.5%. On the other hand, polyvinyl alcohol PVA-based film containing europium-based ternary complexes shows only a slight PCE improvement by ~0.6% [20]. One can easily observe that in either case, the PCE of the solar cells improves, making this approach practically feasible.

The use of phosphor materials as DC coatings for solar cells can be advantageous due to their excellent chemical stability, quantum yields, and quantum-cutting properties. On the other hand, phosphor materials are vulnerable to humidity, which usually quenches luminescence. Hence, the use of hydrophobic coatings like polydimethylsiloxane (PDMS) has several advantages, including optical transparency, water rejection to prevent the luminescence quenching of phosphor materials, and self-cleaning features [21]. To the best of our knowledge, the use of hydrophobic PDMS with embedded luminescent phosphor particles with high quantum yield to improve the PCE of Si solar cells has not yet been reported. Hence, in this study, we incorporated red-emitting $Y_2O_3:Eu^{3+}$ particles into the PDMS matrix to produce a hydrophobic and luminescent coating for solar cells. We found that the developed coating is multifunctional; for example, it can protect the Si solar cells from harmful UV radiation, it has passive radiative cooling, it has self-cleaning properties, and it can also improve the PCE of devices. The hydrophobic properties and structural stability of the produced coating were evaluated by testing it in various types of chemical conditions and under UV irradiation.

2. Materials and Methods

2.1. Synthesis of $Y_2O_3:Eu^{3+}$ Particles

High-purity reagents were purchased from Merck Group (St. Louis, MO, USA) and utilized without any purification. The luminescent $Y_2O_3:Eu^{3+}$ particles were produced using the urea homogeneous precipitation protocol [22,23]. In brief, 0.5 g of urea, 371.5 mg of yttrium nitrate hexahydrate, and 12.8 mg of europium nitrate pentahydrate were completely dissolved in 40 mL of deionized water. The resulting mixture was heated in the oven at 90 °C for 2 h. The obtained white precipitates were collected, dried, and calcined in air at 600 °C for 1 h.

2.2. PDMS-$Y_2O_3:Eu^{3+}$ Coating Deposition

For deposition of the coating with the optimal parameters, 3 mg of the as-prepared $Y_2O_3:Eu^{3+}$ powder was dissolved in 25 µL of hydrophobic agent (1H,1H,2H,2H-perfluorooctyltriethoxysilane) and sonicated for 5 min to form a homogenous dispersed mixture. After the addition of 200 µL of cured PDMS (Sylgard 184 kit, Dow Inc., Midland, MI, USA), the solution was thoroughly mixed, and an additional 25 µL of hydrophobic agent was added. The resulting solution was repeatedly sonicated for ~5–7 min. Finally, the obtained solution was spin-coated on glass slides and Si solar cells at 3000 rpm for 20 s. The coated samples were dried for 24 h at 80–100 °C.

2.3. Characterization

Morphological and elemental examinations were performed using a scanning electron microscope (SEM, Carl Zeiss Auriga Crossbeam 540, Oberkochen, Germany) equipped with energy-dispersive X-ray spectroscopy (EDX, Aztec Oxford Instruments, Abingdon, UK). The photoluminescence analysis, quantum yield, reflectance and absorbance measurements were carried out with a Quantaurus absolute quantum yield spectrometer (C9920-02, Hamamatsu Photonics K.K., Hamamatsu, Japan) equipped with an integrating sphere. X-ray diffraction (XRD) measurements were performed using a SmartLab X-ray Diffractometer (Rigaku Corp., Tokyo, Japan) with a Cu Kα radiation source. Transmittance measurements were performed using the Genesys 50 UV–Visible spectrophotometer (Thermo Fisher Scientific Inc., Waltham, MA, USA). The hydrophobic properties of the films were tested by contact angle goniometer (Ossila Ltd., Sheffield, UK). The current–voltage (J-V) measurements were performed using a semiconductor parameter analyzer (Agilent B1500A, Agilent Technologies Inc., Santa Clara, CA, USA). The samples were illuminated using AAA class Oriel Sol3A solar simulator (Newport-Spectra Physics GmbH, Darmstadt, Germany). An AM 1.5 G filter and Si reference cell were applied to adjust the light intensity.

3. Results and Discussion

The deposition process was optimized through testing of several concentrations of Y_2O_3:Eu^{3+} particles in PDMS and also by varying the thickness of the coatings. However, the experimental data discussion is confined to the optimal coating conditions only. Figure 1 displays the overall scheme and processes that take place when light photons with different energies hit the surface of bare glass or glass with a coating. Some photons will be reflected, while others will pass through and generate electron–hole pairs in the active layer of the solar cell. Down-conversion optical materials have absorption in the UV-blue range and are commonly deposited on top of solar cells, which in turn simplifies the coating process. In this study, the PDMS acts as a hydrophobic matrix that houses the downconversion particles, while these particles convert UV-blue photons into visible light photons. In this study, Y_2O_3:Eu^{3+} particles were selected as a down-conversion optical material because of the synthesis simplicity, large Stokes shift, and high quantum yield [22].

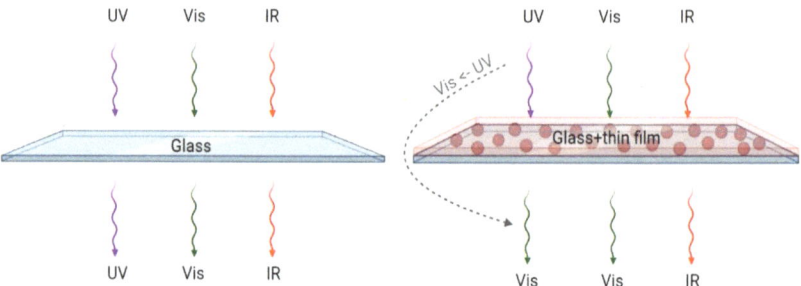

Figure 1. Schematic representation of light conversion by photoluminescent coating.

The morphological examination of the Y_2O_3:Eu^{3+} particles was conducted using TEM and SEM. Figure 2A shows that the produced particles have a spherical shape and range in size from ~350 to 500 nm. Figure S1 (Supporting Information) confirms the even distribution of the key elements in the sample, with yttrium (Y), oxygen (O), and europium (Eu) effectively detected. Figure 2B shows a cross-sectional image of the hydrophobic and luminescent coating taken for the optimal sample. Cross-sectional SEM analysis revealed that the coating thickness was ~5.4 μm for the optimal samples. Furthermore, the successful incorporation of Y_2O_3:Eu^{3+} particles into a PDMS matrix was also validated, with some Y_2O_3:Eu^{3+} particles visible at the border of the coating. The XRD analysis of the prepared Y_2O_3:Eu^{3+} particles (Figure S2, Supporting Information) revealed distinct diffraction peaks at 2θ angles of 20.7, 29.2, 33.8, 35.9, 39.9, 43.5, 48.6, and 57.6, corresponding to the diffraction of the (211), (222), (400), (411), (332), (413), (440), and (622) crystal planes, respectively. The observed XRD pattern suggested that the Y_2O_3:Eu^{3+} particles adopted a body-centered cubic (bcc) phase of Y_2O_3 [24].

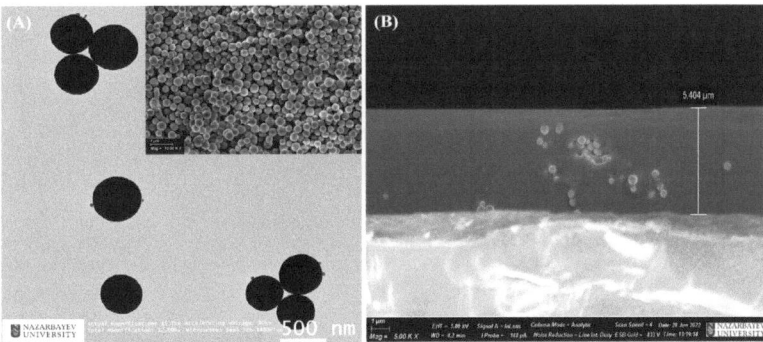

Figure 2. (**A**) TEM and SEM (inset) images of Y_2O_3:Eu^{3+} particles. (**B**) Cross-sectional SEM image of PDMS-Y_2O_3:Eu^{3+} coating.

Figure 3A shows the "Commission Internationale de l'éclairage" CIE chromaticity diagram and excitation/emission spectra of Y_2O_3:Eu^{3+} particles. The emission pattern of Y_2O_3:Eu^{3+} particles is represented by $^5D_0 \rightarrow ^7F_0$, $^5D_0 \rightarrow ^7F_1$, $^5D_0 \rightarrow ^7F_2$, $^5D_0 \rightarrow ^7F_3$ transitions in the yellow–red region, with the $^5D_0 \rightarrow ^7F_2$ electric dipole transition at 612 nm being the most pronounced and dominant. The corresponding transitions are schematically shown in Figure S3 (Supporting Information). The excitation curve (λ_{em} = 612 nm) shows a broad band in the UV region, which is associated with a charge transfer band from the 2p orbital of O^{2-} to the 4f orbital of Eu^{3+} [22,24]. The measured absolute quantum yield QY of Y_2O_3:Eu^{3+} particles was found to be ~78.3%, which is considered to be exceptionally high compared to other organic/inorganic optical materials [25,26]. The CIE diagram also confirmed the successful light conversion from UV to red, with the following emission chromaticity coordinates (x = 0.591; y = 0.330). It should be outlined that the majority of solar cells, including silicon, perovskite, and dye-sensitized have notable light absorption in the visible–near IR regions. Hence, UV to red down-conversion of Y_2O_3:Eu^{3+} particles with high QY can be used to shield solar cells from destructive UV radiation and at the same time improve the efficiency of solar cells by supplying additional photons. In the next step, we tested the light transmittance of the PDMS-Y_2O_3:Eu^{3+} coating with optimal thickness to that of the bare glass slide. Figure 3B shows a small reduction in light transmittance when compared to the reference glass slide. In particular, the light transmittance was reduced by 12.7% (at 320 nm), 11.5% (at 500 nm), and 8.5% (at 800 nm) after deposition of the PDMS-Y_2O_3:Eu^{3+} coating. Figure 3B inset shows that the PDMS- PDMS-Y_2O_3:Eu^{3+} coating with average light transmittance of ~80% or more was visually transparent in white light but glowed red when exposed to UV light.

Figure 4 depicts the absorbance and reflectance studies of the PDMS-Y_2O_3:Eu^{3+} coating measured in an integrating sphere. One can notice that absorbance of the PDMS-Y_2O_3:Eu^{3+} coating was observed in the UV range (~260–400 nm) with a maximum close to 304 nm. Hence, we can speculate that the PDMS-Y_2O_3:Eu^{3+} coating indeed only absorbed UV photons and converted them further into visible light photons. A similar trend was also observed with the reflectance study; a minimal reflection was observed in the region of ~260–400 nm, which corroborated well the absorbance results.

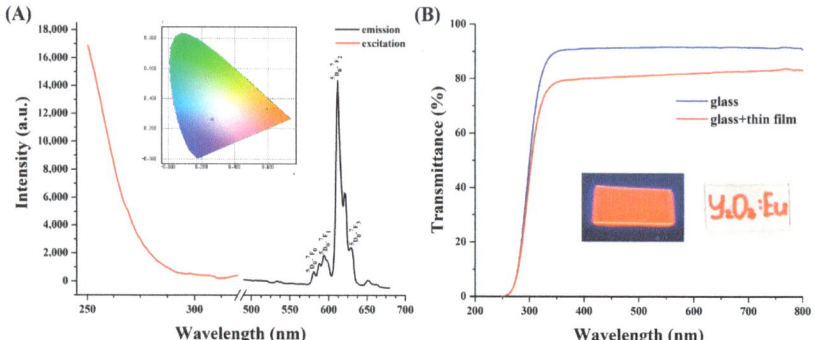

Figure 3. (**A**) PL emission and excitation of Y_2O_3:Eu^{3+} particles. (**B**) Transmittance of a bare glass slide and a glass slide with a PDMS-Y_2O_3:Eu^{3+} coating. Inset are digital images of the PDMS-Y_2O_3:Eu^{3+} coating under UV and daylight illumination.

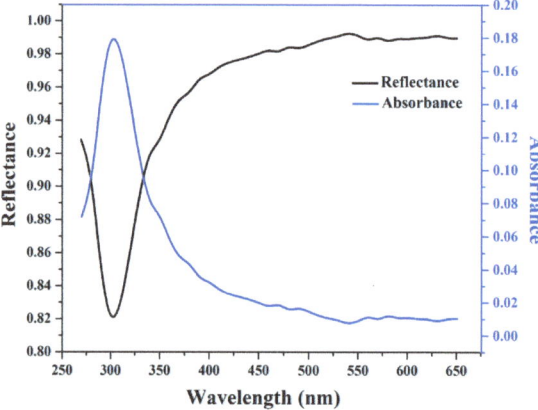

Figure 4. Absorbance and reflectance measurements of PDMS-Y_2O_3:Eu^{3+} coating on glass slide.

The hydrophobicity of the formed PDMS-Y_2O_3:Eu^{3+} coatings was assessed by measuring the contact angles using a goniometer. The optimized samples displayed contact angles ranging from 119 to 121°, indicating their hydrophobic properties. Furthermore, the structural stability of the coating was further assessed by a series of additional experiments. For example, the UV stability was estimated by placing the coated glass slides under an ultraviolet lamp (λ = 365 nm, 8 W) at a distance of 2 cm for 24 h, with a measuring of the contact angle every 3 h. Moreover, the chemical stability of the coating was determined by immersing the coatings in different chemical environments, i.e., pH 3, pH 5, pH 7 (distilled water), and NaCl (1M) solutions for 24 h. Figure 5A shows that after being exposed to UV light, the measured contact angles ranged between 118 and 120°, indicating good resistance of the coating to UV irradiation. It should be outlined that the UV flux reaching the Earth's surface is lower than that of a UV lamp; hence, the developed PDMS-Y_2O_3:Eu^{3+} coatings have the potential to endure long-term sun irradiation. Figure 5B indicates that the contact angles of the coatings also did not vary substantially after being placed in various chemical media simulating acidic conditions. Hence, it can be concluded that the PDMS-Y_2O_3:Eu^{3+} coating may retain hydrophobic properties for an extended period, which is potentially useful for self-cleaning purposes.

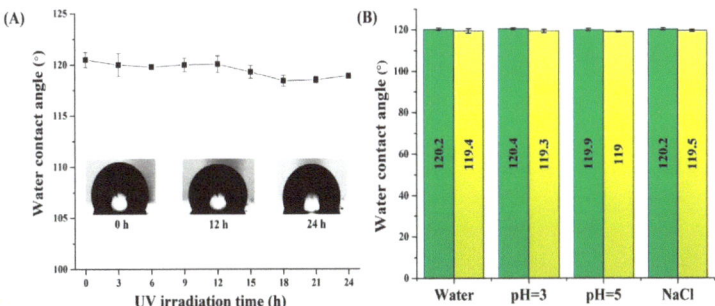

Figure 5. Effect of (**A**) UV irradiation time and (**B**) impact of various media on the hydrophobicity of PDMS-Y_2O_3:Eu^{3+} coatings.

As a proof-of-concept, the photovoltaic (PV) parameters of polycrystalline Si cells (active area = 1 × 1 cm) were tested before and after the deposition of PDMS-Y_2O_3:Eu^{3+} with the optimized thickness. Figure 6A,B illustrate the corresponding J-V curves and external quantum efficiency (EQE) of the Si cells (n = 3) before and after the deposition of the coating, respectively. Figure 7 shows that all key PV parameters were improved; i.e., one can achieve PCE improvement in the Si cells following PDMS-Y_2O_3:Eu^{3+} deposition. On average, the PCE enhancement was found to be ~9.23%. It should be also emphasized that the EQE pattern of the coated sample followed the proposed idea, i.e., to absorb photons in the UV-blue region and convert them to visible photons in the red region. Generally, all key PV changes can be associated with the management of the optical characteristics, and a similar trend was observed in the literature [27,28]. Typically, one can observe that the V_{oc}, J_{sc}, and FF values were improved to ~3.57%, 1.38%, and 10.1% respectively. Among them, the J_{sc} enhancement was associated with increased light-generated current, while the V_{oc} improvement was typically associated with improved concentration or lower recombination rates of charge carriers [29,30]. The FF is frequently linked to parasitic resistive losses [31], and the FF can gradually increase with the growth in the irradiance (in this case, irradiance is increasing in the red and near IR regions), which reaches a maximum point and then decreases [32]. Furthermore, the FF lowers, as the temperature rises [32]; so, the PDMS-Y_2O_3:Eu^{3+} coating also reduces the heating of Si cells similar to a PDMS-SiO_2-based radiative cooling film [33]. For example, under solar illumination, an uncoated Si cell reached approximately 54.9 °C in 5 min, whereas a coated cell reached approximately 52.7 °C, as shown in Figure S4 (Supporting Information). Hence, we can speculate that the overall PCE enhancement is associated with the synergetic combination of light conversion by PDMS-Y_2O_3:Eu^{3+}, improved generation of charge carriers in Si cells, and passive cooling of the PDMS-Y_2O_3:Eu^{3+} coating.

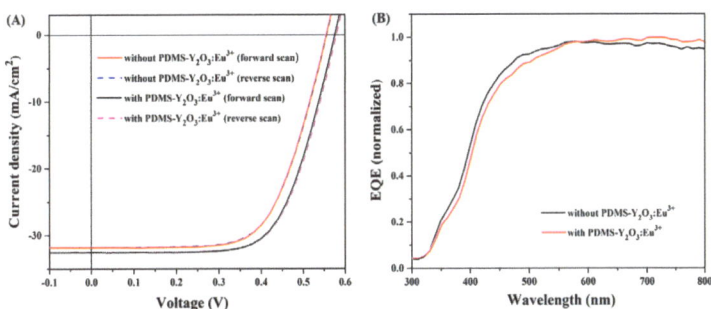

Figure 6. (**A**) J–V and (**B**) EQE characteristics of Si solar cells with and without PDMS-Y_2O_3:Eu^{3+} coatings.

PDMS-Y_2O_3:Eu^{3+} coating	Scan direction	V_{oc} (V)	J_{sc} (mA/cm²)	FF (%)	PCE (%)
without	forward	0.56±0.01	31.96±0.25	59.23±1.09	11.20±0.23
without	reverse	0.56±0.01	32.00±0.26	59.19±1.05	11.19±0.25
with	forward	0.58±0.01	32.43±0.19	65.12±0.64	12.24±0.16
with	reverse	0.58±0.01	32.41±0.18	65.27±0.61	12.22±0.14

Figure 7. Key PV parameters for Si cells before and after the deposition of luminescent PDMS-Y_2O_3:Eu^{3+} coatings.

The self-cleaning property of the prepared PDMS-Y_2O_3:Eu^{3+} coating was further shown on a glass slide (not visible on the Si cells due to the dark surface). Figure 8 shows the digital images of the coated glass slide with soil spots on the surface. Because the adhesion between the surface and dust is weaker than that between the water droplet and dust, the water droplet will clean the surface coating as it rolls down.

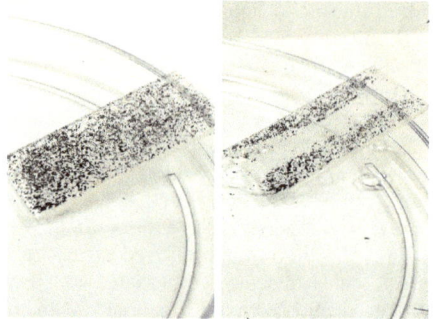

Figure 8. Demonstration of the self-cleaning properties of the PDMS-Y_2O_3:Eu^{3+} coating. The right image depicts the coating after several water droplets have been dropped.

Finally, the UV protection capabilities of the coatings with optimized thicknesses were investigated under the constant UV lamp illumination (λ = 365 nm, 8 W, t = 1 h) of the bare and coated Si cells (n = 3 per batch). We found that, on average, the coated Si cells had a lower PCE drop (~0.8%) as compared to those of the bare Si cells (~1.8%). Hence, it can be concluded that PDMS-Y_2O_3:Eu^{3+} can be potentially employed as a hydrophobic UV-blocking and light-converting coating for the potential power enhancement of Si cells. Si solar cell fabrication is a mature process, with a reported efficiency degradation by several % within a year [34]. As a result, the short-term testing of non-coated and coated Si cells at 80 °C yielded no appreciable decrease in the PCE. Therefore, field trials, such as film deposition using spray coating on large-sized panels, as well as long-term durability testing under temperature, humidity, and light fluctuations will be explored in future studies.

4. Conclusions

In summary, we successfully prepared a hydrophobic and luminescent PDMS-Y_2O_3:Eu^{3+} coating for potential applications in photovoltaic solar cells. The prepared coating contains Y_2O_3:Eu^{3+} particles, dispersed in a hydrophobic PDMS matrix, which convert UV photons to visible (red) photons with a QY of ~78.3%. As a result, down-converted red photons can be partially reabsorbed by Si cells, resulting in a PCE improvement of ~9.23%. The preliminary data on the UV stability, chemical stability, and UV protection suggested that the PDMS-Y_2O_3:Eu^{3+} coating is durable and can be employed on UV-sensitive solar cells for UV protection and power enhancement purposes.

Supplementary Materials: The following supporting information can be downloaded at: https://www.mdpi.com/article/10.3390/nano14080674/s1, Figure S1: EDS elemental mapping of Y_2O_3:Eu^{3+} particles; Figure S2: XRD pattern of Y_2O_3:Eu^{3+} particles; Figure S3: Schematic transitions within the Eu^{3+} ion. Figure S4: Heating of Si cells within 5 min under simulated solar light illumination.

Author Contributions: Conceptualization, T.S.A.; methodology, T.S.A.; validation, D.G., K.Z., S.S. and Z.Y.; formal analysis, D.G., K.Z., S.S., Z.Y., A.N. and T.S.A.; investigation, D.G., K.Z., S.S., Z.Y., A.N. and T.S.A.; resources, A.N. and T.S.A.; data curation, D.G., K.Z., Z.Y. and S.S.; writing—original draft preparation, D.G. and K.Z.; writing—review and editing, A.N. and T.S.A.; supervision, T.S.A.; funding acquisition, T.S.A. All authors have read and agreed to the published version of the manuscript.

Funding: This research was funded by Nazarbayev University FDCRDG (Grant No. 20122022FD4111). This work was also supported by the Sustainability Living Lab (SLL) program at Nazarbayev University in cooperation with the "National Conservation Initiative" funded by Chevron Corporation.

Data Availability Statement: The original contributions presented in the study are included in the article/Supplementary Material; further inquiries can be directed to the corresponding author.

Acknowledgments: The authors thanks Laura Khamkhash for help with SEM measurements.

Conflicts of Interest: The authors declare that this study received funding from Chevron Corporation. The funder was not involved in the study design, collection, analysis, interpretation of data, the writing of this article or the decision to submit it for publication. The authors declare no conflict of interest with a funding received from Nazarbayev University.

References

1. Saga, T. Advances in Crystalline Silicon Solar Cell Technology for Industrial Mass Production. *NPG Asia Mater.* **2010**, *2*, 96–102. [CrossRef]
2. Battaglia, C.; Cuevas, A.; De Wolf, S. High-Efficiency Crystalline Silicon Solar Cells: Status and Perspectives. *Energy Environ. Sci.* **2016**, *9*, 1552–1576. [CrossRef]
3. Andreani, L.C.; Bozzola, A.; Kowalczewski, P.; Liscidini, M.; Redorici, L. Silicon Solar Cells: Toward the Efficiency Limits. *Adv. Phys. X* **2018**, *4*, 1548305. [CrossRef]
4. Addie, A.J.; Ismail, R.A.; Mohammed, M.A. Amorphous Carbon Nitride Dual-Function Anti-Reflection Coating for Crystalline Silicon Solar Cells. *Sci. Rep.* **2022**, *12*, 9902. [CrossRef] [PubMed]
5. Spence, M.; Hammond, R.; Pockett, A.; Wei, Z.; Johnson, A.; Watson, T.; Carnie, M.J. A Comparison of Different Textured and Non-Textured Anti-Reflective Coatings for Planar Monolithic Silicon-Perovskite Tandem Solar Cells. *ACS Appl. Energy Mater.* **2022**, *5*, 5974–5982. [CrossRef]
6. Lindroos, J.; Savin, H. Review of Light-Induced Degradation in Crystalline Silicon Solar Cells. *Sol. Energy Mater. Sol. Cells* **2016**, *147*, 115–126. [CrossRef]
7. Perrakis, G.; Tasolamprou, A.C.; Kenanakis, G.; Economou, E.N.; Tzortzakis, S.; Kafesaki, M. Ultraviolet Radiation Impact on the Efficiency of Commercial Crystalline Silicon-Based Photovoltaics: A Theoretical Thermal-Electrical Study in Realistic Device Architectures. *OSA Contin.* **2020**, *3*, 1436. [CrossRef]
8. Sinha, A.; Qian, J.; Moffitt, S.L.; Hurst, K.; Terwilliger, K.; Miller, D.C.; Schelhas, L.T.; Hacke, P. UV-induced Degradation of High-efficiency Silicon PV Modules with Different Cell Architectures. *Prog. Photovolt. Res. Appl.* **2022**, *31*, 36–51. [CrossRef]
9. Johansson, W.; Peralta, A.; Jonson, B.; Anand, S.; Österlund, L.; Karlsson, S. Transparent TiO_2 and ZnO Thin Films on Glass for UV Protection of PV Modules. *Front. Mater.* **2019**, *6*, 259. [CrossRef]
10. Yousefi, F.; Mousavi, S.B.; Heris, S.Z.; Naghash-Hamed, S. UV-Shielding Properties of a Cost-Effective Hybrid PMMA-Based Thin Film Coatings Using TiO_2 and ZnO Nanoparticles: A Comprehensive Evaluation. *Sci. Rep.* **2023**, *13*, 7116. [CrossRef]
11. Hong, M.; Xuan, T.; Liu, J.; Jiang, Z.; Chen, Y.; Chen, X.; Li, H. Air-Exposing Microwave-Assisted Synthesis of CuInS2/ZnS Quantum Dots for Silicon Solar Cells with Enhanced Photovoltaic Performance. *RSC Adv.* **2015**, *5*, 102682–102688. [CrossRef]
12. Xuan, T.-T.; Liu, J.-Q.; Li, H.-L.; Sun, H.-C.; Pan, L.; Chen, X.-H.; Sun, Z. Microwave Synthesis of High Luminescent Aqueous CdSe/CdS/ZnS Quantum Dots for Crystalline Silicon Solar Cells with Enhanced Photovoltaic Performance. *RSC Adv.* **2015**, *5*, 7673–7678. [CrossRef]
13. Du, P.; Lim, J.H.; Leem, J.W.; Cha, S.M.; Yu, J.S. Enhanced Photovoltaic Performance of Dye-Sensitized Solar Cells by Efficient Near-Infrared Sunlight Harvesting Using Upconverting Y_2O_3:Er^{3+}/Yb^{3+} Phosphor Nanoparticles. *Nanoscale Res. Lett.* **2015**, *10*, 321. [CrossRef] [PubMed]
14. Lee, S.; Kim, C.U.; Bae, S.; Liu, Y.; Noh, Y.I.; Zhou, Z.; Leu, P.W.; Choi, K.J.; Lee, J. Improving Light Absorption in a Perovskite/Si Tandem Solar Cell via Light Scattering and UV-Down Shifting by a Mixture of SiO_2 Nanoparticles and Phosphors. *Adv. Funct. Mater.* **2022**, *32*, 2204328. [CrossRef]

15. Sekar, R.; Ravitchandiran, A.; Angaiah, S. Recent Advances and Challenges in Light Conversion Phosphor Materials for Third-Generation Quantum-Dot-Sensitized Photovoltaics. *ACS Omega* **2022**, *7*, 35351–35360. [CrossRef]
16. Trupke, T.; Green, M.A.; Würfel, P. Improving Solar Cell Efficiencies by Down-Conversion of High-Energy Photons. *J. Appl. Phys.* **2002**, *92*, 1668–1674. [CrossRef]
17. Cui, J.; Li, P.; Chen, Z.; Cao, K.; Li, D.; Han, J.; Shen, Y.; Peng, M.; Fu, Y.Q.; Wang, M. Phosphor Coated NiO-Based Planar Inverted Organometallic Halide Perovskite Solar Cells with Enhanced Efficiency and Stability. *Appl. Phys. Lett.* **2016**, *109*, 171103. [CrossRef]
18. Lopez-Delgado, R.; Zhou, Y.; Zazueta-Raynaud, A.; Zhao, H.; Pelayo, J.E.; Vomiero, A.; Álvarez-Ramos, M.E.; Rosei, F.; Ayon, A. Enhanced Conversion Efficiency in Si Solar Cells Employing Photoluminescent Down-Shifting CdSe/CdS Core/Shell Quantum Dots. *Sci. Rep.* **2017**, *7*, 14104. [CrossRef]
19. Zhou, D.; Liu, D.; Pan, G.; Chen, X.; Li, D.; Xu, W.; Bai, X.; Song, H. Cerium and Ytterbium Codoped Halide Perovskite Quantum Dots: A Novel and Efficient Downconverter for Improving the Performance of Silicon Solar Cells. *Adv. Mater.* **2017**, *29*, 1704149. [CrossRef]
20. Yang, D.; Liang, H.; Liu, Y.; Hou, M.; Kan, L.; Yang, Y.; Zang, Z. A Large-Area Luminescent Downshifting Layer Containing an Eu^{3+} Complex for Crystalline Silicon Solar Cells. *Dalton Trans.* **2020**, *49*, 4725–4731. [CrossRef]
21. Yang, J.W.; Kim, D.I.; Jeong, R.H.; Park, S.; Boo, J. Enhancement of Perovskite Solar Cell Performance by External Down-conversion of Eu-complex Film. *Int. J. Energy Res.* **2022**, *46*, 7996–8006. [CrossRef]
22. Atabaev, T.S.; Thi Vu, H.H.; Kim, H.-K.; Hwang, Y.-H. The Optical Properties of Eu^{3+} and Tm^{3+} Codoped Y_2O_3 Submicron Particles. *J. Alloys Compd.* **2012**, *525*, 8–13. [CrossRef]
23. Atabaev, T.S.; Shin, Y.; Song, S.-J.; Han, D.-W.; Hong, N. Toxicity and T_2-Weighted Magnetic Resonance Imaging Potentials of Holmium Oxide Nanoparticles. *Nanomaterials* **2017**, *7*, 216. [CrossRef]
24. Chávez-García, D.; Sengar, P.; Juárez-Moreno, K.; Flores, D.L.; Calderón, I.; Barrera, J.; Hirata, G.A. Luminescence Properties and Cell Uptake Analysis of Y_2O_3:Eu, Bi Nanophosphors for Bio-Imaging Applications. *J. Mater. Res. Technol.* **2021**, *10*, 797–807. [CrossRef]
25. Yang, J.; Fang, M.; Li, Z. Organic Luminescent Materials: The Concentration on Aggregates from Aggregation-induced Emission. *Aggregate* **2020**, *1*, 6–18. [CrossRef]
26. Fang, M.-H.; Bao, Z.; Huang, W.-T.; Liu, R.-S. Evolutionary Generation of Phosphor Materials and Their Progress in Future Applications for Light-Emitting Diodes. *Chem. Rev.* **2022**, *122*, 11474–11513. [CrossRef]
27. Thi Vu, H.H.; Atabaev, T.S.; Ahn, J.Y.; Dinh, N.N.; Kim, H.-K.; Hwang, Y.-H. Dye-Sensitized Solar Cells Composed of Photoactive Composite Photoelectrodes with Enhanced Solar Energy Conversion Efficiency. *J. Mater. Chem. A* **2015**, *3*, 11130–11136. [CrossRef]
28. Minemoto, T.; Mizuta, T.; Takakura, H.; Hamakawa, Y. Antireflective Coating Fabricated by Chemical Deposition of ZnO for Spherical Si Solar Cells. *Sol. Energy Mater. Sol. Cells* **2007**, *91*, 191–194. [CrossRef]
29. Deibel, C. Photocurrent Generation in Organic Solar Cells. *Semicond. Semimet.* **2011**, *85*, 297–330. [CrossRef]
30. He, Y.; Liu, J.; Sung, S.-J.; Chang, C. Downshifting and Antireflective Thin Films for Solar Module Power Enhancement. *Mater. Des.* **2021**, *201*, 109454. [CrossRef]
31. Lin, H.; Wang, G.; Su, Q.; Han, C.; Xue, C.; Yin, S.; Fang, L.; Xu, X.; Gao, P. Unveiling the Mechanism of Attaining High Fill Factor in Silicon Solar Cells. *Prog. Photovolt. Res. Appl.* **2024**, in press. [CrossRef]
32. Saloux, E.; Teyssedou, A.; Sorin, M. Explicit Model of Photovoltaic Panels to Determine Voltages and Currents at the Maximum Power Point. *Sol. Energy* **2011**, *85*, 713–722. [CrossRef]
33. Tu, Y.; Tan, X.; Yang, X.; Qi, G.; Yan, K.; Kang, Z. Antireflection and Radiative Cooling Difunctional Coating Design for Silicon Solar Cells. *Opt. Express* **2023**, *31*, 22296. [CrossRef] [PubMed]
34. Pascual, J.; Martinez-Moreno, F.; García, M.; Marcos, J.; Marroyo, L.; Lorenzo, E. Long-term Degradation Rate of Crystalline Silicon PV Modules at Commercial PV Plants: An 82-MWp Assessment over 10 Years. *Prog. Photovolt. Res. Appl.* **2021**, *29*, 1294–1302. [CrossRef]

Disclaimer/Publisher's Note: The statements, opinions and data contained in all publications are solely those of the individual author(s) and contributor(s) and not of MDPI and/or the editor(s). MDPI and/or the editor(s) disclaim responsibility for any injury to people or property resulting from any ideas, methods, instructions or products referred to in the content.

Article

Enhancing the Performance of Nanocrystalline SnO₂ for Solar Cells through Photonic Curing Using Impedance Spectroscopy Analysis

Moulay Ahmed Slimani, Jaime A. Benavides-Guerrero, Sylvain G. Cloutier and Ricardo Izquierdo *

Département de Génie Électrique, École de Technologie Supérieure, 1100 Rue Notre-Dame Ouest, Montréal, QC H3C 1K3, Canada; moulay-ahmed.slimani.1@ens.etsmtl.ca (M.A.S.); jaime-alberto.benavides-guerrero.1@ens.etsmtl.ca (J.A.B.-G.); sylvaing.cloutier@etsmtl.ca (S.G.C.)
* Correspondence: ricardo.izquierdo@etsmtl.ca

Abstract: Wide-bandgap tin oxide (SnO₂) thin-films are frequently used as an electron-transporting layers in perovskite solar cells due to their superior thermal and environmental stabilities. However, its crystallization by conventional thermal methods typically requires high temperatures and long periods of time. These post-processing conditions severely limit the choice of substrates and reduce the large-scale manufacturing capabilities. This work describes the intense-pulsed-light-induced crystallization of SnO₂ thin-films using only 500 µs of exposure time. The thin-films' properties are investigated using both impedance spectroscopy and photoconductivity characteristic measurements. A Nyquist plot analysis establishes that the process parameters have a significant impact on the electronic and ionic behaviors of the SnO₂ films. Most importantly, we demonstrate that light-induced crystallization yields improved topography and excellent electrical properties through enhanced charge transfer, improved interfacial morphology, and better ohmic contact compared to thermally annealed (TA) SnO₂ films.

Keywords: impedance spectroscopy; photonic curing; SnO₂; dark injection current transient; photo-Celiv

1. Introduction

Electron-transporting layers (ETLs) are critical components in most optoelectronic device architectures, including perovskite solar cells (PSCs). These PSC devices rely on organic–inorganic perovskite materials to efficiently absorb light and generate charge carriers [1–3]. ETL layers are essential for promoting efficient electron transport, block holes, align energy levels, and ultimately enhance the efficiency and stability of perovskite solar cells. Choosing appropriate ETL materials is essential for the performance of PSCs. Typical ETL materials require processing between 150 and 500 °C, resulting in higher processing times and energy costs. Most importantly, this prevents their integration on most low-cost substrates that require processing temperatures below 150 °C [4,5]. In this context, intense pulsed light annealing, also sometimes referred to as photonic curing (PC) [6], is an emerging technique that is ideally suited for large-scale manufacturing as is relies on short, high-intensity light pulses to anneal materials selectively and rapidly [7,8]. In this process, the optical energy absorbed by the active material can sustain carefully controlled light-induced annealing with minimal substrate damage. As a result, even metals with relatively high melting points can be successfully sintered on low-cost plastic- or paper-based substrates [9–11]. As such, this technique is also especially well-suited for roll-to-roll (R2R) manufacturing [12]. SnO₂ metal-oxide thin-films were first utilized as ETLs for perovskite-based solar cells nearly a decade ago [13,14]. They have since emerged as the preferred material for PSCs over TiO₂ and ZnO due to their large band gaps, higher charge mobilities, and better stabilities under ambient conditions [15–17]. A few years later, SnO$_2$ films were photonically annealed in just 20 ms, enabling the fabrication of PSCs with reduced hysteresis and a 15% power conversion efficiency [9]. However, these previous studies did not address the effect of photonic curing on the electronic properties of SnO$_2$ films. To investigate this, we used impedance spectroscopy (IS), which is a rapid technique for evaluating these properties. IS is a powerful tool to shed light on the kinetic processes taking place within electrochemical systems [18,19]. During measurement, a small alternating current (AC) signal is coupled with a direct current (DC) voltage and is applied to the device. The phase difference between the DC voltage and AC current is measured over a wide frequency range to identify the various physical effects in the device. As a result, IS measurements can assess the physical and chemical processes of various types of devices, including optoelectronic devices, fuel cells, and solid-state batteries [20]. IS is a non-destructive [14,21,22] tool that can be effectively used to optimize the stability and performance of these devices by characterizing their charge transport properties [18,23]. Typically, the IS measurements exhibit two arcs corresponding to low-frequency (LF) and high-frequency (HF) responses, respectively [24,25]. The series resistance (Rs), charge-transfer resistance (R_{CT}), and parallel capacitance can be determined from the HF and LF responses.

This work explores the impact of the photonic curing parameters on thin-film SnO$_2$ properties using IS and photocurrent characteristic analysis to unveil and control the ionic and electronic kinetics within the treated SnO$_2$ layer. As we demonstrate, this improved understanding and control leads to enhanced electronic properties with great potential for improved perovskite solar cell manufacturability.

2. Experimental Section

Commercial patterned fluorine-doped tin Oxide (FTO) substrates (Shenzhen Huayu Union Technology, Shenzhen, China, resistance: 7 Ohm/sq) doped with fluorine are cleaned using a sequential process of 10 min each in an ultrasonic bath with DI water, acetone, and isopropyl alcohol (IPA). After drying with a nitrogen spray gun, residual organic contaminants are removed by performing a 15 min O$_2$ plasma treatment (Plasma Etch, Carson City, NV, USA, PE-100LF). To prepare the SnO$_2$ solution, a colloidal precursor of SnO$_2$ obtained from Alfa Aesar (15% in H$_2$O colloidal dispersion CN: 044592.A3) is diluted with DI water to a concentration of 3% by volume. The SnO$_2$ solution is spin-coated onto the clean FTO substrate in one step in air 3000 rpm for 30 s. The edges of the FTO electrodes are then cleaned with a dry cotton swab to enable electrical and IS measurements (Figure 1). For TA, SnO$_2$ films are annealed using a hot plate at 150 °C for 30 min under ambient air. For photonic curing, each sample is treated using a Novacentrix PulseForge system (500 V/3 A) power supply with 3 capacitors providing radiant energy greater than 20 J.cm^{-2} using a lamp system (7.6 cm × 60.8 cm) with an illumination area of 300 mm × 75 mm. The light source ensures uniform curing over a large area and delivers short (20 µs to 100 ms) but intense light pulses from a broadband xenon flash lamp (200–1500 nm). A Paois (Fluxim AG, SN:20121 Winterthur, Switzerland) tool is used for all electrical and IS measurements. SEM (SU8230 Hitachi) and AFM (Bruker, MultiMode8, Billerica, MA, USA) are used for topography inspection. For impedance spectroscopy, the FTO edges that are used as electrodes are connected to the Paois to measure the impedance over a range of frequencies (10 Hz to 10 MHz) in the dark at 0.07 V perturbation at room temperature. Impedance data can be analyzed using Nyquist and Bode plots to interpret electrochemical properties such as the charge transfer resistance, capacitance, and dielectric properties. The temperature is simulated using NovaCentrix SimPulse software, this simulation package is standard on the PulseForge Version 3, Austin, TX, USA. The configuration is modeled as follows (from bottom to top): aluminum chuck, 6 mm; glass, 2.2 mm; and FTO, 600 nm. The thicknesses of the glass and FTO layers are taken from the manufacturers. X-ray diffraction (XRD) is done using a Bruker D8 Advance (Billerica, MA, USA), and optical absorbance is done using a UV–Vis–NIR spectrophotometer from Perkin Elmer (Waltham, MA, USA).

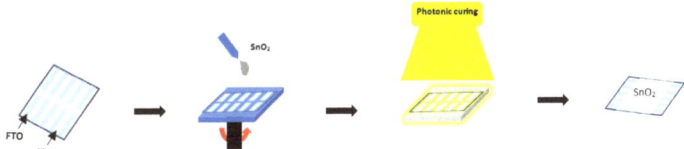

Figure 1. Illustration of the SnO$_2$ sample fabrication process.

3. Results and Discussion

After deposition of colloidal SnO$_2$ films using the protocol, samples are post-processed using varying pulse durations and energy densities using the methodology described in the Experimental Section. To investigate the impact of PC on the electrical properties of SnO$_2$ films, we conduct flash annealing for pulse durations of 500, 1500, 2500, and 3500 µs, followed by photocurrent measurements. This allows us to optimize our photonic annealing parameters and define the high photoconductivity range for SnO$_2$ films. Photocurrent analysis is used to map the different zones' photoconductivity. Pulses ranging from 500 to 3500 µs are utilized to complete the photo-responsivity characterization. Figure 2a shows the I–V responses in the dark and under illumination for two samples photonically treated using a pulse duration of 2500 µs and, respectively, 2 J.cm^{-2} and 4 J.cm^{-2}. A low photo-responsivity indicates that the illumination and dark curves approach the overlap limit, while a high photo-responsivity indicates a clear offset (more than 0.5 order of magnitude) between the I–V characteristics in the dark and under illumination. Based on such measurements, Figure 2b displays a photo-responsivity map for samples photonically treated using different pulse durations vs. energy densities. To shed light on these results, IS and SEM characterizations are conducted. SnO$_2$ is highly transparent, which makes photonic curing difficult [26]. To mitigate this problem, we use substrates with FTO patterns that act as a structural support and a stable base for the growth of SnO$_2$ nanoparticles. This helps promote the transmission of the heat generated when light is absorbed by the nanoparticles [27], which can increase the local temperature around the nanoparticles and promote the recrystallization process. FTO substrates exhibit rougher surfaces than glass [28], promoting superior adhesion and growth of SnO$_2$ nanoparticles [29]. Their conductivity enhances the electrical properties of the resulting SnO$_2$ films. The FTO substrate's roughness directly influences both the diameter and alignment of the SnO$_2$ nanoparticles [30]. Areas with FTO patterns acting as a blanket allow for changes in nanoparticle recrystallization depending on the energy density used.

Figure 2c displays SEM images of the bare FTO substrate, and Figure 2d–f show SnO$_2$ films deposited on FTO and photonically treated using energy densities of 0.15, 2.06, and 2.46 J.cm^{-2}, respectively. As the energy density is increased from 0.15 to 2.06 to 2.46 J.cm^{-2} while using 1500 µs pulse durations, the SnO$_2$-covered films appear increasingly granular, while the distinct grain boundaries that were clearly observed in the FTO/glass film are less apparent. The recrystallization of the SnO$_2$ film follows the substrate topography well, revealing the underlying FTO grain profile. This process indicates that higher energy densities lead to improved film–substrate adhesion and more pronounced exposure of the underlying grain structure. The photonic curing of SnO$_2$ wet films enables water evaporation and subsequent crystallization of SnO$_2$ nanoparticles [31]. The degree of crystallization greatly affects the photoconductivity of SnO$_2$ films and their ability to carry charge carriers [32,33]. A film's properties largely depend on two independent parameters: the energy density and the pulse duration of the pulsed light.

To obtain quantitative information and to better understand the surface morphology and roughness, we also conduct AFM analyses on samples subjected to different types of annealing treatments. Figure 2g shows the surface roughness of the film samples for a scan area of 5 × 5 µm^2. It highlights the improvement in surface topography after optimal photonic treatment with SnO$_2$, with a root-mean-square roughness of 14.01 nm, compared to 45.57 nm for the thermally annealed sample. Roughness is defined as the microscopic and macroscopic variations on a material's surface [34]. It measures the irregularities present on the surface. These variations can have a significant impact on the physical and chemical properties of materials, such as the recombination rates in ETL films for solar cells. Low-roughness films can reduce the recombination rate and thus improve performance [35]. The morphological effect of PC processing can be beneficial in terms of device performance. This significant improvement underscores another important advantage of photonic treatment for enhancing the quality of SnO_2 as an electron transport layer (ETL) in perovskite solar cells.

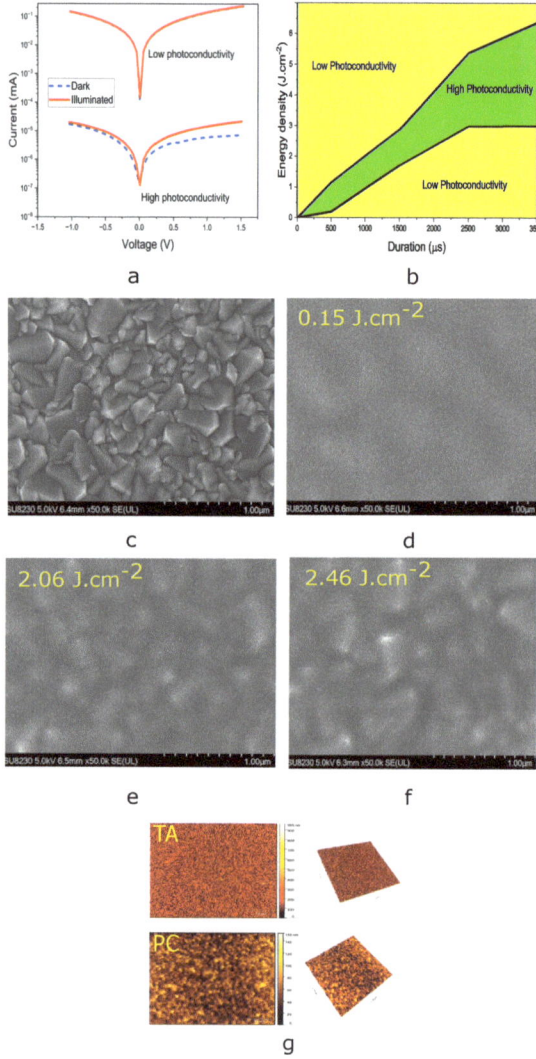

Figure 2. (a) I–V responses in the dark and under illumination for two samples photonically treated using a pulse duration of 2500 µs and, respectively, 2 and 4 J.cm^{-2}. (b) Photo-response map for samples photonically treated using different pulse durations vs. energy densities based on the criterion in Figure 2a. (c) SEM images of FTO/glass. (d–f) SEM images of PC of SnO_2 samples on FTO/glass. (g) Atomic force microscopy (AFM) images in 2D and 3D of thermally and photonically annealed samples.

This section focuses on the variation of IS results for SnO_2 films treated with different energy densities and pulse durations of 500, 1500, 2500, and 3500 µs. For these measurements, the SnO_2 film is deposited onto FTO glass, and its electrochemical behavior can be represented by an equivalent circuit that produces a semicircle on the Nyquist diagram. Figure 3a–d displays IS results for SnO_2 samples treated using these different pulse durations and energy densities. When the pulse duration is fixed and the energy density is increased, the semicircle decreases until it reaches its minimum, and then the arc widens. The frequency response exhibits two distinct behaviors. At high frequencies (HF), it is dominated by the resistance attributed to electronic transport (R_{CT}). At low frequencies (L_F), it is dominated by the recombination resistance (R_{rec}) related to ionic diffusion and charge accumulation at the contacts [36,37]. In Figure 3, it corresponds to the second semicircle inclined at 45° to the real axis in the Nyquist graph [38]. The semi-circle in the high-frequency region is generally related to the counter-electrode and its interface [39]. A smaller half-circle suggests a lower R_{CT} and better photoconductivity of the device. These Nyquist plots suggest that our devices' equivalent circuits can be accurately modeled by a resistor–capacitor (RC) pair in the dark AC regime [40]. As such, the interface contribution can be derived from the equivalent circuit's parameters [41]. The series resistance (Rs) can be obtained by measuring the shift of the semi-circle from the origin along the horizontal axis [42]. However, the time constant related to the physical phenomena dominating at both the low and high frequencies is described by $\tau_{HF} \cdot \omega_{HF} = 1$ and $\tau_{LF} \cdot \omega_{LF} = 1$, with $\omega_{HF,LF} = 2\pi \cdot f_{max,HF,LF}$ [40]. The time constants can be deduced from the IS results by identifying the peak of the semicircle, which corresponds to the maximum frequency, or by calculating $\tau = Req \cdot Ceq$, as shown in Table 1.

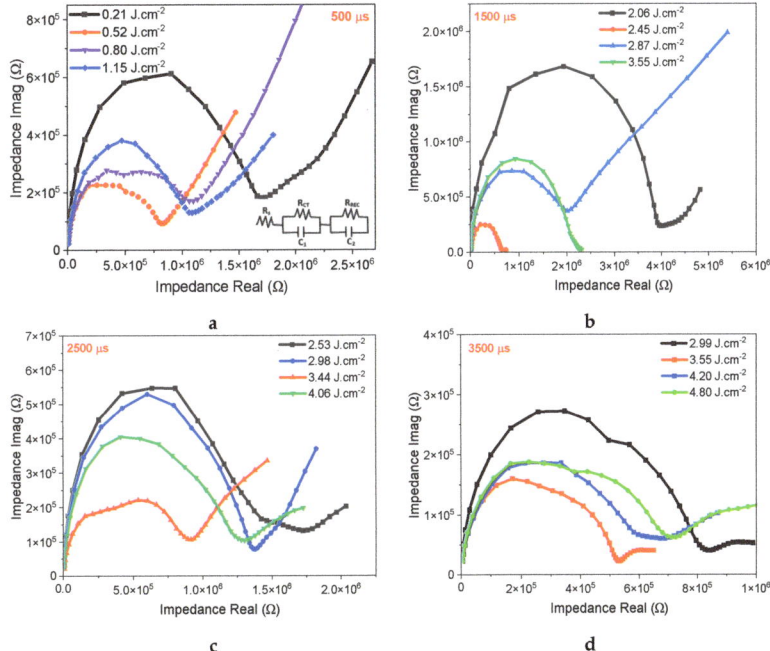

Figure 3. Imaginary versus real components of impedance for photonically annealed films with pulse durations of 500, 1500, 2500, and 3500 µs, respectively.

Table 1. IS parameters extracted from the Nyquist plots for thermally annealed and photonically treated samples at 0 V in dark conditions with 0.07 V perturbations. Photonic treatment is performed using a 3500 µs pulse duration at 3.55 J.cm^{-2} energy density.

Device	Rs (kΩ)	R_{CT} (MΩ)	Ceq (pF)	τ_{HF} (µs)
Thermally annealed	1.96	0.99	0.88	0.87
Photonically treated	3.06	0.49	0.78	0.38

Figure 4a compares the Cole–Cole plots for films that are photonically treated using 500, 1500, 2500, and 3500 µs pulses with respective energy densities of 0.52, 2.45, 3.44, and 3.55 J.cm^{-2} with a typical film sample crystallized using standard thermal annealing. Clearly, the physical and chemical properties of the resulting SnO$_2$ films appear greatly affected by the pulse duration and energy density. When the pulse duration is 3500 µs and the energy density is 3.55 J.cm^{-2}, the high-frequency arc is smallest, suggesting that the film is less resistive and facilitating charge transfer. In comparison, the thermally annealed sample exhibits a larger semicircle than all of the photonically treated samples. This suggests increased imaginary impedance associated with a decrease in charge transfer. Figure 4b–d compare the imaginary impedance, capacitance, and conductance versus the frequency for the best thermally annealed and the best photonically treated films for the conditions 3.55 J.cm^{-2} and 3500 µs. In Figure 4b, the high-frequency (HF) peaks appear between 10^5–10^6 Hz for both samples. The response time can be obtained by taking the inverse of the peak frequency from the imaginary impedance graph. Table 1 presents the IS parameters extracted from the spectra. There, the R_{CT} value for the thermally annealed sample is roughly twice the value achieved using optimal photonic curing conditions. This suggests that the SnO$_2$/FTO interface provides a low R_{CT} under the effect of photonic annealing, which facilitates charge carrier transport. The resulting time constant is 0.8 µs for the thermally annealed film, compared to 0.38 µs for the optimal photonic curing conditions. This suggest that photonically induced crystallization promotes a faster response time, resulting in low recombination and more dominant ionic diffusion behavior [43,44]. At low frequencies, the thermally annealed device does not exhibit any measurable peak, which is consistent with the presence of the single semicircle in Figure 4b. In contrast, the impedance plot of the photonically treated device is curved at low frequencies, explaining the start of the second semicircle in this region. Frequency, time constant, and conductivity values are good indicators of process kinetics [45,46]. Indeed, the dark IS can be directly related to the carrier density, mobility, and conductivity [38]. The temperature simulation results using the photonic annealing parameters shown in Figure 4e reveal a relationship between the energy density, pulse duration, and resulting temperature of the SnO$_2$ film. As the energy density increases from 0.52 to 3.55 J.cm^{-2}, the temperature increases from 122 to 364 °C then decreases to 329 °C for the film treated with an energy density of 3.55 J.cm^{-2} and a pulse duration of 3500 µs. These parameters are crucial to determine the energy transferred to the SnO$_2$ film, but they show a non-linear trend with temperature. Figure 4f shows X-ray diffraction (XRD) measurements of the thermally and photonically annealed SnO$_2$ films. The prominent peaks are determined to correspond to (110), (101), (200), (211), (220), and (002), confirming the tetragonal crystal structure of SnO$_2$ for both the TA and PC films [47–49].

Figure 4c,d show capacitance and conductivity evolutions as a function of the operation frequency. Figure 4c illustrates two distinct capacitance behaviors, each corresponding to a specific polarization process. This distinction makes it possible to identify specific capacitive processes directly from the plot [50,51]. The high-frequency capacitance C_{HF} (above 100 kHz) exhibits a plateau in the order of 1 pF for both thermally and photonically treated devices and is rather similar for both annealing processes. This region represents the geometric capacitance and is due to the intrinsic dielectric polarization of the SnO$_2$ layer [50]. However, photonic treatment achieves higher capacitance values at low frequencies (below 1 kHz) compared to the thermally annealed device. This is primarily due to the accumulation of charges or ions [52,53] resulting from the polarization of the interfaces between the SnO$_2$ layer and the electrodes. At low frequencies, the increase in capacitance is dominated by ionic movement in the dark and electronic movement in the light [54,55]. In circuits that exhibit capacitive behavior, the capacitor offers less resistance to the flow of alternating current as the frequency increases. Accordingly, Figure 4d shows increases in conductance for both devices in the high-frequency region. This behavior is consistent with that of semiconductors, where capacitance and conductance vary inversely [56–58].

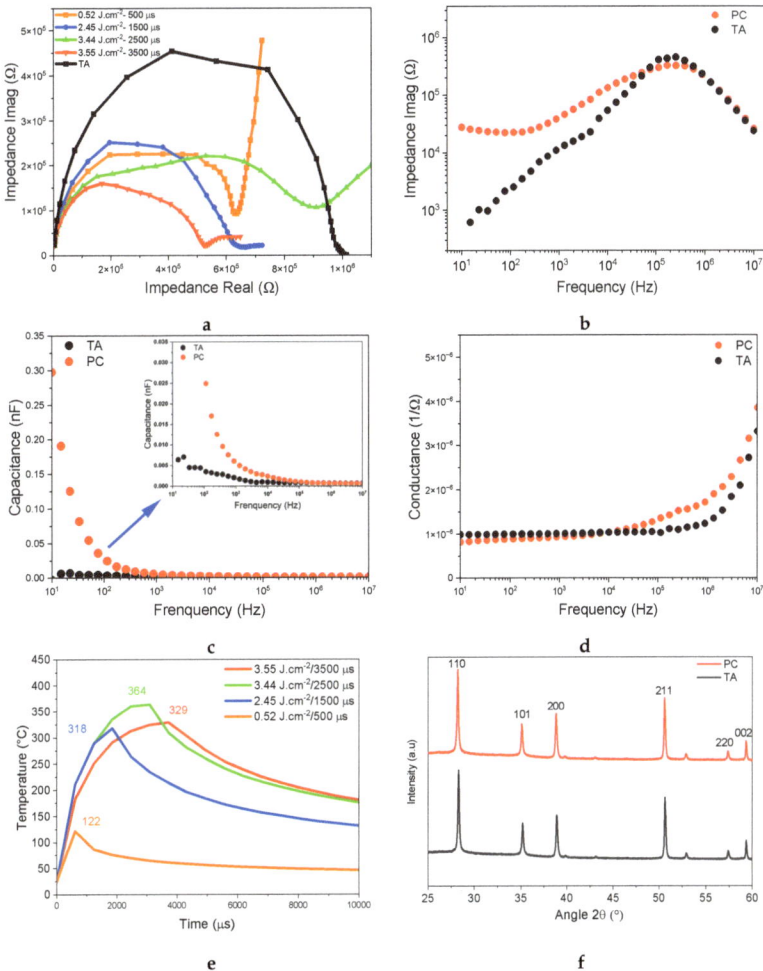

Figure 4. (**a**) Cole–Cole plot for films thermally and photonically treated using 500, 1500, 2500, and 3500 µs with energy densities of 0.52, 2.45, 3.44, and 3.55 J.cm^{-2}, respectively. (**b**–**d**) Comparison of imaginary impedance, capacitance, and conductance vs. frequency for typical thermally annealed and photonically treated samples. (**e**) SimPulse simulations of temperature profiles of photonically annealed SnO$_2$ film Cole–Cole plots for films photonically treated using 500, 1500, 2500, and 3500 µs with energy densities of 0.52, 2.45, 3.44, and 3.55 J.cm^{-2}, respectively. (**f**) XRD spectra of thermally and photonically annealed SnO$_2$ films at 3.55 J.cm^{-2} and 3500 µs.

The optical properties of the prepared samples are characterized by UV–Vis absorption spectra. As shown in Figure 5a, the transmittance of PC-treated films is higher than that of TA-treated films, which is desirable for solar cell applications. SnO$_2$ is a direct bandgap (BG) semiconductor; its BG can be calculated using a Tauc plot [59], as shown in Figure 5b. The calculated BGs are 3.45 eV and 3.43 eV for the TA- and PC-treated films, respectively, which explains why the TA film is slightly more transparent than the PC film. Measurements in Figure 5c,d compare the dark injection transients for the photocurrent rise and decay for the thermally and photonically treated (3.55 J.cm^{-2}, 3500 μs) samples. This time-of-flight technique is useful for determining majority carrier mobility and trapping, especially in thin-films [60]. Figure 5c illustrates that the current for the photonically treated film rises to 2.7 mA, compared to 2.3 mA for the thermally annealed film. The current also increases more rapidly in the photonically treated sample, reflecting the interrelationship between charge carrier generation and recombination. Therefore, the rapid increase in current for the PC sample can be attributed to the fast accumulation of photogenerated carriers [61]. Figure 5d compares the decay of the transient current. After reaching its maximum, the current decay depends on the charge capture coefficient [62]. The decay graph illustrates the speed of charge recombination after being excited by a 1.2 V pulse voltage. A shorter carrier lifetime suggests faster recombination and a high carrier capture rate, which implies more rapid current decay for the thermally annealed sample. In contrast, photonic curing yields a lower recombination rate, resulting in slower decay and longer current holding times. The photogeneration and recombination processes have a significant impact on the density and mobility of charge carriers. Figure 5e compares the charge mobility using the photo-CELIV technique using the following expression [63–65]:

$$\mu = \frac{2d^2}{3A.t_{max}^2(1 + 0.36\frac{\Delta}{J_{max}})} \quad (1)$$

where d is the SnO$_2$ film thickness, A is the slope of the extraction voltage ramp, t_{max} is the time related to the current peak, and Δ is the difference between the maximum current and the displacement current plateau. Photo-CELIV is a technique used to extract the charge mobility by illuminating the device. The measurement displays the current overshoot and the time at which the current reaches its maximum, which is an essential parameter for quantifying mobility. However, it should be noted that Photo-CELIV only measures fast carriers and cannot distinguish between the mobility of electrons and holes. The Photo-CELIV measurements for the film after optimized photonic treatment yield 4.56×10^{-2} V cm^2 s^{-1}, compared with 3.66×10^{-2} V cm^2 s^{-1} for the thermally annealed film. This measurement does not precisely reflect the mobility of the SnO$_2$ material. However, it serves as a characterization for comparing the fastest or maximum carrier mobility values. This higher maximum mobility compared to thermal annealing is consistent with previous results.

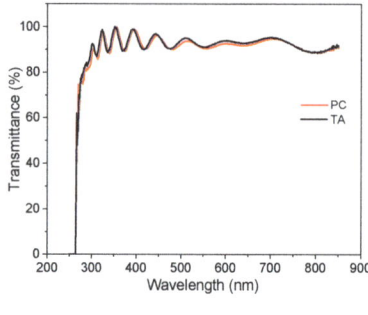

a

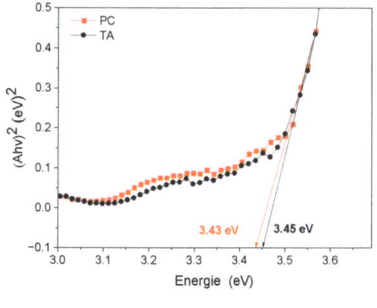

b

Figure 5. *Cont.*

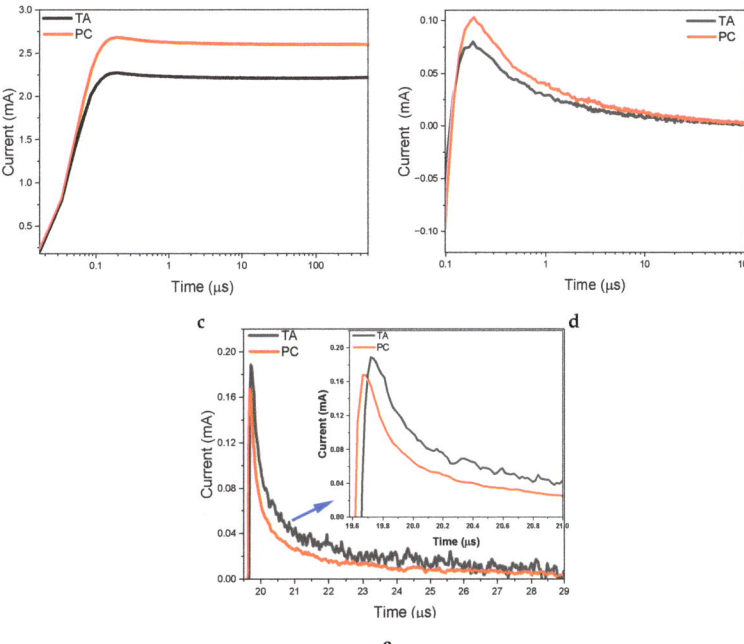

Figure 5. (**a,b**) Transmittance spectra and Tauc plots of thermally and photonically annealed SnO_2 samples. (**c,d**) Dark injection transients for the photocurrent rise and decay for the thermally and photonically treated samples. (**e**) Charge mobility using the photo-CELIV technique for the thermally and photonically treated samples.

4. Conclusions

In summary, we propose an optimized photonic annealing approach to improve the electrical properties of SnO_2 thin-films compared to standard annealing. SnO_2 thin-films play an essential role in emerging device architectures, especially as the electron-transporting layer (ETL) for perovskite-based solar cells. We use impedance spectroscopy to analyze the electrical behavior of SnO_2 films in the dark. The results indicate that the impedance spectroscopy response depends significantly on both the energy density and the pulse duration and shed light on the resulting ionic and electronic transfer. Additionally, we demonstrate that photonic treatment yields SnO_2 layers with enhanced electrical performance and a significantly reduced manufacturing time compared to standard thermal annealing. This would be a great advantage for large-scale manufacturing of better and cheaper perovskite-based solar cells.

Author Contributions: Writing—original draft, methodology, resources, conceptualization, and formal analysis, M.A.S.; simulation and manuscript review, J.A.B.-G.; supervision, visualization, and review and editing, S.G.C. and R.I. All authors have read and agreed to the published version of the manuscript.

Funding: The authors would like to acknowledge the financial support received from the NSERC via Discovery grants RGPIN-2023-05211 and RGPIN 2022-03083 as well as from the Canada Research Chairs (CRC-2021-00490).

Data Availability Statement: The data that support the findings of this study are available from the corresponding author upon reasonable request.

Conflicts of Interest: The authors declare no conflicts of interest.

Abbreviations

The following abbreviations are used in this manuscript:

ETL Electron-transporting layer
PSC Perovskite solar cell
TA   Thermal annealing
PC   Photonic curing

References

1. Wei, Z.; Zhao, Y.; Jiang, J.; Yan, W.; Feng, Y.; Ma, J. Research progress on hybrid organic–inorganic perovskites for photo-applications. *Chin. Chem. Lett.* **2020**, *31*, 3055–3064. [CrossRef]
2. Jing, H.; Zhu, Y.; Peng, R.W.; Li, C.Y.; Xiong, B.; Wang, Z.; Liu, Y.; Wang, M. Hybrid organic-inorganic perovskite metamaterial for light trapping and photon-to-electron conversion. *Nanophotonics* **2020**, *9*, 3323–3333. [CrossRef]
3. Yang, Z.; Lai, J.; Zhu, R.; Tan, J.; Luo, Y.; Ye, S. Electronic Disorder Dominates the Charge-Carrier Dynamics in Two-Dimensional/Three-Dimensional Organic–Inorganic Perovskite Heterostructure. *J. Phys. Chem. C* **2022**, *126*, 12689–12695. [CrossRef]
4. Jiang, Q.; Chu, Z.; Wang, P.; Yang, X.; Liu, H.; Wang, Y.; Yin, Z.; Wu, J.; Zhang, X.; You, J. Planar-structure perovskite solar cells with efficiency beyond 21%. *Adv. Mater.* **2017**, *29*, 1703852. [CrossRef]
5. Liu, C.; Hu, M.; Zhou, X.; Wu, J.; Zhang, L.; Kong, W.; Li, X.; Zhao, X.; Dai, S.; Xu, B. Efficiency and stability enhancement of perovskite solar cells by introducing CsPbI3 quantum dots as an interface engineering layer. *NPG Asia Mater.* **2018**, *10*, 552–561. [CrossRef]
6. Schroder, K.A. Mechanisms of photonic curing™: Processing high temperature films on low temperature substrates. *Nanotechnology* **2011**, *2*, 220–223.
7. Akhavan, V.; Schroder, K.; Farnsworth, S. Photonic Curing. *Inkjet Print. Ind. Mater. Technol. Syst. Appl.* **2022**, *2*, 1051–1064.
8. Secor, E.B.; Ahn, B.Y.; Gao, T.Z.; Lewis, J.A.; Hersam, M.C. Rapid and versatile photonic annealing of graphene inks for flexible printed electronics. *Adv. Mater.* **2015**, *27*, 6683–6688. [CrossRef]
9. Zhu, M.; Liu, W.; Ke, W.; Clark, S.; Secor, E.B.; Song, T.B.; Kanatzidis, M.G.; Li, X.; Hersam, M.C. Millisecond-pulsed photonically-annealed tin oxide electron transport layers for efficient perovskite solar cells. *J. Mater. Chem. A* **2017**, *5*, 24110–24115. [CrossRef]
10. Altay, B.N.; Turkani, V.S.; Pekarovicova, A.; Fleming, P.D.; Atashbar, M.Z.; Bolduc, M.; Cloutier, S.G. One-step photonic curing of screen-printed conductive Ni flake electrodes for use in flexible electronics. *Sci. Rep.* **2021**, *11*, 3393. [CrossRef]
11. Piper, R.T.; Daunis, T.B.; Xu, W.; Schroder, K.A.; Hsu, J.W. Photonic curing of nickel oxide transport layer and perovskite active layer for flexible perovskite solar cells: A path towards high-throughput manufacturing. *Front. Energy Res.* **2021**, *9*, 640960. [CrossRef]
12. Maskey, B.B.; Koirala, G.R.; Kim, Y.; Park, H.; Yadav, P.; Park, J.; Sun, J.; Cho, G. Photonic Curing for Enhancing the Performance of Roll-to-Roll Printed Electronic Devices. Oklahoma State University, USA. 2019. Available online: https://core.ac.uk/reader/215230370 (accessed on 31 July 2024).
13. Dong, Q.; Shi, Y.; Wang, K.; Li, Y.; Wang, S.; Zhang, H.; Xing, Y.; Du, Y.; Bai, X.; Ma, T. Insight into perovskite solar cells based on SnO2 compact electron-selective layer. *J. Phys. Chem. C* **2015**, *119*, 10212–10217. [CrossRef]
14. Matacena, I.; Guerriero, P.; Lancellotti, L.; Alfano, B.; De Maria, A.; La Ferrara, V.; Mercaldo, L.V.; Miglietta, M.L.; Polichetti, T.; Rametta, G. Impedance spectroscopy analysis of perovskite solar cell stability. *Energies* **2023**, *16*, 4951. [CrossRef]
15. Hu, M.; Zhang, L.; She, S.; Wu, J.; Zhou, X.; Li, X.; Wang, D.; Miao, J.; Mi, G.; Chen, H. Electron transporting bilayer of SnO2 and TiO2 nanocolloid enables highly efficient planar perovskite solar cells. *Sol. RRL* **2020**, *4*, 1900331. [CrossRef]
16. Irfan, M.; Ünlü, F.; Lê, K.; Fischer, T.; Ullah, H.; Mathur, S. Electrospun Networks of ZnO-SnO2 Composite Nanowires as Electron Transport Materials for Perovskite Solar Cells. *J. Nanomater.* **2022**, *2022*, 6043406. [CrossRef]
17. Martínez-Denegri, G.; Colodrero, S.; Kramarenko, M.; Martorell, J. All-nanoparticle SnO2/TiO2 electron-transporting layers processed at low temperature for efficient thin-film perovskite solar cells. *ACS Appl. Energy Mater.* **2018**, *1*, 5548–5556. [CrossRef]
18. Magar, H.S.; Hassan, R.Y.; Mulchandani, A. Electrochemical impedance spectroscopy (EIS): Principles, construction, and biosensing applications. *Sensors* **2021**, *21*, 6578. [CrossRef]
19. Pascoe, A.R.; Duffy, N.W.; Scully, A.D.; Huang, F.; Cheng, Y.B. Insights into planar CH3NH3PbI3 perovskite solar cells using impedance spectroscopy. *J. Phys. Chem. C* **2015**, *119*, 4444–4453. [CrossRef]
20. Sinclair, D.C. Characterisation of electro-materials using ac impedance spectroscopy. *Boletín Soc. Española Cerámica Vidr.* **1995**, *34*, 55–65.
21. Middlemiss, L.A.; Rennie, A.J.; Sayers, R.; West, A.R. Characterisation of batteries by electrochemical impedance spectroscopy. *Energy Rep.* **2020**, *6*, 232–241. [CrossRef]
22. Shohan, S.; Harm, J.; Hasan, M.; Starly, B.; Shirwaiker, R. Non-destructive quality monitoring of 3D printed tissue scaffolds via dielectric impedance spectroscopy and supervised machine learning. *Procedia Manuf.* **2021**, *53*, 636–643. [CrossRef]
23. Cherian, C.T.; Zheng, M.; Reddy, M.; Chowdari, B.; Sow, C.H. Zn2SnO4 nanowires versus nanoplates: Electrochemical performance and morphological evolution during Li-cycling. *ACS Appl. Mater. Interfaces* **2013**, *5*, 6054–6060. [CrossRef] [PubMed]

24. Suo, Z.; Xiao, Z.; Li, S.; Liu, J.; Xin, Y.; Meng, L.; Liang, H.; Kan, B.; Yao, Z.; Li, C.; et al. Efficient and stable inverted structure organic solar cells utilizing surface-modified SnO_2 as the electron transport layer. *Nano Energy* **2023**, *118*, 109032. [CrossRef]
25. Dıaz-Flores, L.; Ramırez-Bon, R.; Mendoza-Galvan, A.; Prokhorov, E.; Gonzalez-Hernandez, J. Impedance spectroscopy studies on SnO_2 films prepared by the sol–gel process. *J. Phys. Chem. Solids* **2003**, *64*, 1037–1042. [CrossRef]
26. Piper, R.T.; Xu, W.; Hsu, J.W. How Optical and Electrical Properties of ITO Coated Willow Glass Affect Photonic Curing Outcome for Upscaling Perovskite Solar Cell Manufacturing. *IEEE J. Photovoltaics* **2022**, *12*, 722–727. [CrossRef]
27. Albrecht, A.; Rivadeneyra, A.; Abdellah, A.; Lugli, P.; Salmerón, J.F. Inkjet printing and photonic sintering of silver and copper oxide nanoparticles for ultra-low-cost conductive patterns. *J. Mater. Chem. C* **2016**, *4*, 3546–3554. [CrossRef]
28. Pan, D.; Fan, H.; Li, Z.; Wang, S.; Huang, Y.; Jiao, Y.; Yao, H. Influence of substrate on structural properties and photocatalytic activity of TiO_2 films. *Micro Nano Lett.* **2017**, *12*, 82–86. [CrossRef]
29. Hamdi, M.; Saleh, M.N.; Poulis, J.A. Improving the adhesion strength of polymers: Effect of surface treatments. *J. Adhes. Sci. Technol.* **2020**, *34*, 1853–1870. [CrossRef]
30. Bandara, T.; Aththanayake, A.; Kumara, G.; Samarasekara, P.; DeSilva, L.A.; Tennakone, K. Transparent and conductive F-Doped SnO_2 nanostructured thin films by sequential nebulizer spray pyrolysis. *MRS Adv.* **2021**, *6*, 417–421. [CrossRef]
31. Ghahremani, A.H.; Martin, B.; Gupta, A.; Bahadur, J.; Ankireddy, K.; Druffel, T. Rapid fabrication of perovskite solar cells through intense pulse light annealing of SnO_2 and triple cation perovskite thin films. *Mater. Des.* **2020**, *185*, 108237. [CrossRef]
32. Mukhamedshina, D.M.; Beisenkhanov, N.B. Influence of crystallization on the properties of SnO_2 thin films. In *Advances in Crystallization Processes*; IntechOpen: London, UK, 2012.
33. Zhu, Z.; Bai, Y.; Liu, X.; Chueh, C.; Yang, S.; Jen, A.K. Enhanced efficiency and stability of inverted perovskite solar cells using highly crystalline SnO_2 nanocrystals as the robust electron-transporting layer. *Adv. Mater.* **2016**, *28*, 6478–6484. [CrossRef] [PubMed]
34. Creager, S.E.; Hockett, L.A.; Rowe, G.K. Consequences of microscopic surface roughness for molecular self-assembly. *Langmuir* **1992**, *8*, 854–861. [CrossRef]
35. Keshtmand, R.; Zamani-Meymian, M.R.; Mohamadkhani, F.; Taghavinia, N. Smoothing and coverage improvement of SnO_2 electron transporting layer by NH_4F treatment: Enhanced fill factor and efficiency of perovskite solar cells. *Sol. Energy* **2021**, *228*, 253–262. [CrossRef]
36. Salado, M.; Contreras-Bernal, L.; Caliò, L.; Todinova, A.; López-Santos, C.; Ahmad, S.; Borras, A.; Idígoras, J.; Anta, J.A. Impact of moisture on efficiency-determining electronic processes in perovskite solar cells. *J. Mater. Chem. A* **2017**, *5*, 10917–10927. [CrossRef]
37. Li, N.; Yan, J.; Ai, Y.; Jiang, E.; Lin, L.; Shou, C.; Yan, B.; Sheng, J.; Ye, J. A low-temperature TiO_2/SnO_2 electron transport layer for high-performance planar perovskite solar cells. *Sci. China Mater* **2020**, *63*, 207–215. [CrossRef]
38. Shibuya, H.; Inoue, S.; Ihara, M. Evaluation of dye-sensitized solar cells using forward bias applied impedance spectroscopy under dark. *ECS Trans.* **2009**, *16*, 93. [CrossRef]
39. Abdulrahim, S.M.; Ahmad, Z.; Bahadra, J.; Al-Thani, N.J. Electrochemical impedance spectroscopy analysis of hole transporting material free mesoporous and planar perovskite solar cells. *Nanomaterials* **2020**, *10*, 1635. [CrossRef]
40. Bredar, A.R.; Chown, A.L.; Burton, A.R.; Farnum, B.H. Electrochemical impedance spectroscopy of metal oxide electrodes for energy applications. *ACS Appl. Energy Mater.* **2020**, *3*, 66–98. [CrossRef]
41. Chang, B.Y.; Park, S.M. Integrated description of electrode/electrolyte interfaces based on equivalent circuits and its verification using impedance measurements. *Anal. Chem.* **2006**, *78*, 1052–1060. [CrossRef]
42. Matacena, I. Equivalent circuit extraction procedure from Nyquist plots for graphene-silicon solar cells. In Proceedings of the 2019 15th Conference on Ph.D Research in Microelectronics and Electronics (PRIME), Lausanne, Switzerland, 15–18 July 2019; pp. 273–276. [CrossRef]
43. Prochowicz, D.; Trivedi, S.; Parikh, N.; Saliba, M.; Kalam, A.; Mahdi Tavakoli, M.; Yadav, P. In the Quest of Low-Frequency Impedance Spectra of Efficient Perovskite Solar Cells. *Energy Technol.* **2021**, *9*, 2100229. [CrossRef]
44. Alvarez, A.O.; Arcas, R.; Aranda, C.A.; Bethencourt, L.; Mas-Marzá, E.; Saliba, M.; Fabregat-Santiago, F. Negative capacitance and inverted hysteresis: Matching features in perovskite solar cells. *J. Phys. Chem. Lett.* **2020**, *11*, 8417–8423. [PubMed]
45. Laschuk, N.O.; Easton, E.B.; Zenkina, O.V. Reducing the resistance for the use of electrochemical impedance spectroscopy analysis in materials chemistry. *RSC Adv.* **2021**, *11*, 27925–27936. [PubMed]
46. Hernández, H.H.; Reynoso, A.R.; González, J.T.; Morán, C.G.; Hernández, J.M.; Ruiz, A.M.; Hernández, J.M.; Cruz, R.O. Electrochemical impedance spectroscopy (EIS): A review study of basic aspects of the corrosion mechanism applied to steels. In *Electrochemical Impedance Spectroscopy*; IntechOpen: London, UK, 2020; pp. 137–144.
47. Deva Arun Kumar, K.; Valanarasu, S.; Capelle, A.; Nar, S.; Karim, W.; Stolz, A.; Aspe, B.; Semmar, N. Nanostructured Oxide (SnO_2, FTO) Thin Films for Energy Harvesting: A Significant Increase in Thermoelectric Power at Low Temperature. *Micromachines* **2024**, *15*, 188. [CrossRef] [PubMed]
48. Patil, G.E.; Kajale, D.D.; Gaikwad, V.B.; Jain, G.H. Preparation and characterization of SnO_2 nanoparticles by hydrothermal route. *Int. Nano Lett.* **2012**, *2*, 17. [CrossRef]
49. Peiris, T.N.; Benitez, J.; Sutherland, L.; Sharma, M.; Michalska, M.; Scully, A.D.; Vak, D.; Gao, M.; Weerasinghe, H.C.; Jasieniak, J. A stable aqueous SnO_2 nanoparticle dispersion for roll-to-Roll fabrication of flexible perovskite solar cells. *Coatings* **2022**, *12*, 1948. [CrossRef]

50. Guerrero, A.; Garcia-Belmonte, G.; Mora-Sero, I.; Bisquert, J.; Kang, Y.S.; Jacobsson, T.J.; Correa-Baena, J.P.; Hagfeldt, A. Properties of contact and bulk impedances in hybrid lead halide perovskite solar cells including inductive loop elements. *J. Phys. Chem. C* **2016**, *120*, 8023–8032. [CrossRef]
51. Almora, O.; Zarazua, I.; Mas-Marza, E.; Mora-Sero, I.; Bisquert, J.; Garcia-Belmonte, G. Capacitive dark currents, hysteresis, and electrode polarization in lead halide perovskite solar cells. *J. Phys. Chem. Lett.* **2015**, *6*, 1645–1652. [CrossRef] [PubMed]
52. Mahapatra, A.; Parikh, N.; Kumar, P.; Kumar, M.; Prochowicz, D.; Kalam, A.; Tavakoli, M.M.; Yadav, P. Changes in the electrical characteristics of perovskite solar cells with aging time. *Molecules* **2020**, *25*, 2299. [CrossRef] [PubMed]
53. Todinova, A.; Contreras-Bernal, L.; Salado, M.; Ahmad, S.; Morillo, N.; Idígoras, J.; Anta, J.A. Towards a universal approach for the analysis of impedance spectra of perovskite solar cells: Equivalent circuits and empirical analysis. *ChemElectroChem* **2017**, *4*, 2891–2901. [CrossRef]
54. Zarazua, I.; Han, G.; Boix, P.P.; Mhaisalkar, S.; Fabregat-Santiago, F.; Mora-Seró, I.; Bisquert, J.; Garcia-Belmonte, G. Surface recombination and collection efficiency in perovskite solar cells from impedance analysis. *J. Phys. Chem. Lett.* **2016**, *7*, 5105–5113. [CrossRef]
55. Zarazua, I.; Bisquert, J.; Garcia-Belmonte, G. Light-induced space-charge accumulation zone as photovoltaic mechanism in perovskite solar cells. *J. Phys. Chem. Lett.* **2016**, *7*, 525–528. [CrossRef] [PubMed]
56. Li, J.V.; Ferrari, G. *Capacitance Spectroscopy of Semiconductors*; CRC Press: Boca Raton, FL, USA, 2018.
57. Heeger, A.J.; MacDiarmid, A.G.; Shirakawa, H. *The Nobel Prize in Chemistry, 2000: Conductive Polymers*; Royal Swedish Academy of Sciences: Stockholm, Sweden, 2000; pp. 1–16.
58. Namsheer, K.; Rout, C.S. Conducting polymers: A comprehensive review on recent advances in synthesis, properties and applications. *RSC Adv.* **2021**, *11*, 5659–5697.
59. Wang, Y.; Su, N.; Liu, J.; Lin, Y.; Wang, J.; Guo, X.; Zhang, Y.; Qin, Z.; Liu, J.; Zhang, C.; et al. Enhanced visible-light photocatalytic properties of SnO_2 quantum dots by niobium modification. *Results Phys.* **2022**, *37*, 105515. [CrossRef]
60. Esward, T.; Knox, S.; Jones, H.; Brewer, P.; Murphy, C.; Wright, L.; Williams, J. A metrology perspective on the dark injection transient current method for charge mobility determination in organic semiconductors. *J. Appl. Phys.* **2011**, *109*. [CrossRef]
61. Sarda, N.; Vidhan, A.; Basak, S.; Hazra, P.; Behera, T.; Ghosh, S.; Choudhary, R.J.; Chowdhury, A.; Sarkar, S.K. Photonically Cured Solution-Processed SnO_2 Thin Films for High-Efficiency and Stable Perovskite Solar Cells and Minimodules. *ACS Appl. Energy Mater.* **2023**, *6*, 3996–4006. [CrossRef]
62. Knapp, E.; Ruhstaller, B. The role of shallow traps in dynamic characterization of organic semiconductor devices. *J. Appl. Phys.* **2012**, *112*, 024519. [CrossRef]
63. Aukštuolis, A.; Girtan, M.; Mousdis, G.A.; Mallet, R.; Socol, M.; Rasheed, M.; Stanculescu, A. Measurement of charge carrier mobility in perovskite nanowire films by photo-CELIV method. *Proc. Rom. Acad.-Ser. A Math. Phys. Tech. Sci. Inf. Sci.* **2017**, *18*, 34–41.
64. Stephen, M.; Genevičius, K.; Juška, G.; Arlauskas, K.; Hiorns, R.C. Charge transport and its characterization using photo-CELIV in bulk heterojunction solar cells. *Polym. Int.* **2017**, *66*, 13–25. [CrossRef]
65. Sen, S.; Islam, R. Investigation of Charge Carrier Transport of Bulk Heterojunction Organic Solar Cell Using Photo-CELIV Electrical Simulation. *Proc. Natl. Acad. Sci. India Sect. A Phys. Sci.* **2022**, *92*, 713–717. [CrossRef]

Disclaimer/Publisher's Note: The statements, opinions and data contained in all publications are solely those of the individual author(s) and contributor(s) and not of MDPI and/or the editor(s). MDPI and/or the editor(s) disclaim responsibility for any injury to people or property resulting from any ideas, methods, instructions or products referred to in the content.

Article

Novel Synthesis Route of Plasmonic CuS Quantum Dots as Efficient Co-Catalysts to TiO$_2$/Ti for Light-Assisted Water Splitting

Larissa Chaperman [1], Samiha Chaguetmi [2], Bingbing Deng [1], Sarra Gam-Derrouich [1], Sophie Nowak [1], Fayna Mammeri [1] and Souad Ammar [1,*]

[1] Université Paris Cité, CNRS UMR-7086, ITODYS, 75205 Paris, France; larissa.chaperman@univ-paris-diderot.fr (L.C.); bingbing.deng@univ-paris-diderot.fr (B.D.); sarra.derouich@u-paris.fr (S.G.-D.); sophie.nowak@u-paris.fr (S.N.); fayna.mammeri@u-paris.fr (F.M.)

[2] Faculté des Sciences, Université 20-Août-1955-Skikda, Skikda 21000, Algeria; s.chaguetmi@univ-skikda.dz

* Correspondence: souad.ammar-merah@u-paris.fr

Abstract: Self-doped CuS nanoparticles (NPs) were successfully synthesized via microwave-assisted polyol process to act as co-catalysts to TiO$_2$ nanofiber (NF)-based photoanodes to achieve higher photocurrents on visible light-assisted water electrolysis. The strategy adopted to perform the copper cation sulfidation in polyol allowed us to overcome the challenges associated with the copper cation reactivity and particle size control. The impregnation of the CuS NPs on TiO$_2$ NFs synthesized via hydrothermal corrosion of a metallic Ti support resulted in composites with increased visible and near-infrared light absorption compared to the pristine support. This allows an improved overall efficiency of water oxidation (and consequently hydrogen generation at the Pt counter electrode) in passive electrolyte (pH = 7) even at 0 V bias. These low-cost and easy-to-achieve composite materials represent a promising alternative to those involving highly toxic co-catalysts.

Keywords: plasmonic semiconductive nanoparticles; polyol process; titania photoanodes; water splitting

Citation: Chaperman, L.; Chaguetmi, S.; Deng, B.; Gam-Derrouich, S.; Nowak, S.; Mammeri, F.; Ammar, S. Novel Synthesis Route of Plasmonic CuS Quantum Dots as Efficient Co-Catalysts to TiO$_2$/Ti for Light-Assisted Water Splitting. *Nanomaterials* **2024**, *14*, 1581. https://doi.org/10.3390/nano14191581

Academic Editor: Zhixing Gan

Received: 20 August 2024
Revised: 20 September 2024
Accepted: 27 September 2024
Published: 30 September 2024

Copyright: © 2024 by the authors. Licensee MDPI, Basel, Switzerland. This article is an open access article distributed under the terms and conditions of the Creative Commons Attribution (CC BY) license (https://creativecommons.org/licenses/by/4.0/).

1. Introduction

The quest for sustainable and efficient energy solutions has led to significant interest in the development of advanced materials for water splitting applications. Among these, semiconductive nanostructures play a crucial role in the photoelectrochemical (PEC) water decomposition reactions, which harnesses solar energy to produce hydrogen fuel. TiO$_2$ is among the most prolific materials for light-assisted water electrolysis thanks to its excellent chemical stability and strong photooxidative capabilities, with a band structure triggering water redox potentials [1–3]. Nevertheless, TiO$_2$ presents significant challenges due to its wide bandgap (approximately 3.2 eV for anatase), which limits its light absorption to the UV region, a small fraction of the solar spectrum. Moreover, TiO$_2$ suffers from the rapid recombination of photogenerated electron-hole pairs, which severely impacts its photocatalytic efficiency, limiting electron transfer from the TiO$_2$ conduction band (CB) to the external circuit and then to the cathode materials to be scavenged for proton reduction and hydrogen generation.

A wide variety of strategies was developed to overcome these limitations, the elaboration of titania-based semiconductive hetero-nanostructures being the object of particularly intensive research [4,5]. In such studies, TiO$_2$-based structures are combined to one (or more) sensitizer, a semiconductor with fine-tuned properties that allow to compensate for the TiO$_2$ shortcomings and improve the overall efficiency of the resulting system. The chosen semiconductor must possess appropriate electronic band structures with small bandgap energy and a high absorption coefficient in order to harvest sufficient solar photons. In addition, the energetic levels of its band structure need to straddle the redox potential of the desired water redox reactions, satisfying both the thermodynamics and kinetics requirements for conducting efficient photocatalytic reactions. Furthermore, chemical robustness and photostability are also essential for the selected semiconductors to be considered for such applications to allow for long-term, stable activity in photocatalytic processes [4]. Finally, such materials must be of low toxicity to avoid any deleterious effects on human operators as well as any environmental contamination.

The photocatalytic efficiency of the resulting semiconductive hetero-nanostructures is highly dependent on both the intrinsic photon absorption capability and charge transfer dynamics of the two (or more) photoanode components. Light absorption and charge transfer capabilities are mostly inherent to the band gap of the selected materials and their doping, including self-doping. The narrower the band gap, the higher the photogenerated charge density. As the doping rate increases, the conductivity increases. So, in a standard photoelecrochemical (PEC) cell, in ideal conditions, upon illumination, the photogenerated hole carriers are transported from the bulk of the semiconductor to its electrolyte interface, where they may participate in a water oxidation reaction. Conversely, the photogenerated electron carriers are transported from the bulk of the semiconductor to its titanium interface to be collected through the external circuit by the cathode counterpart to be finally involved in proton reduction into hydrogen.

According to the generalized Marcus theory [6–8], the driving force of electron transfer from a donor to an acceptor (in this case, the CB of the semiconductor to the CB of titania) is determined by the difference in their energetic levels. The logarithm of the rate of charge transfer is defined by a quadratic function with respect to the term of charge transfer driving force [4]. By enlarging the energetic difference between these energies, interfacial charge transfer can be boosted.

By aggregating all these requirements, it becomes evident why metal chalcogenides are particularly well-suited for forming semiconductive hetero-nanostructures with TiO_2 [9], especially when the metal chalcogenide nanostructures consist of nanometal sulfides. Most nanocrystalline metal sulfide compounds exhibit remarkable visible light responsiveness, possess a sufficient number of active sites, and have appropriate reduction and/or oxidation potentials to function as effective photocatalysts [10]. Additionally, their quantum size effects allow for tunable properties such as rapid charge transfer and extended excited-state lifetimes [11].

Among the various combinations with TiO_2, cadmium sulfide (CdS) nanocrystals have been the most extensively studied [12–18], despite concerns about their acute toxicity [19–21]. Our group, along with several others worldwide, has already prepared CdS-TiO_2 composites. We demonstrated that replacing pristine TiO_2 with CdS-TiO_2 as the photoanode in a standard photoelectrochemical (PEC) cell significantly enhances photocurrent generation, even at 0 V bias [12,22].

In practice, by using a controlled hydrothermal corrosion process followed by air calcination at 400 °C of metallic titanium sheets [1], we successfully produced TiO_2/Ti substrates. The resulting titania consisted of interconnected nanofibers (NFs), each about tens of nanometers in diameter and several hundreds of nanometers in length, uniformly covering the metal surface. These nanofibers crystallized in the anatase phase, exhibiting strong UV-range absorption. Using metallic Ti plates as both the support and the precursor for TiO_2 NFs allowed us to streamline the production process while ensuring optimal contact between the photoactive and conductive elements of the photoanode, minimizing current losses [1,23–25].

By selecting a heating temperature below 500 °C, we were able to promote the formation of the anatase phase, which is more suitable for our intended application compared to other titania allotropes [26,27]. Additionally, preformed CdS nanoparticles (NPs), approximately 2–3 nm in size, were deposited onto and between the TiO_2 NFs through a simple ethanol-based impregnation method. This process resulted in a valuable CdS-TiO_2/Ti composite architecture for the targeted application [12,22]. Impregnation is considered as the simplest and most sustainable approach for constructing semiconductive hetero-nanostructures [28–30].

Despite these promising photoelectrochemical (PEC) results, replacing CdS NPs with less toxic yet equally effective metal sulfide NPs nanoparticles remained a key objective of our work. Our goal was to reduce risks to workers, consumers, and the environment during production and handling, ensuring that neither safety nor innovation is compromised, with sustainability playing a central role in our approach.

In this context, we chose to test CuS-TiO$_2$/Ti using a similar simple, low-cost material processing method. Copper sulfide compounds, such as Cu$_2$S and Cu$_7$S$_4$, are well known for their unique optical and electrical properties. Specifically, CuS has a narrow band gap with an energy range between 2.0 and 2.2 eV [31,32]. It can exhibit either p-type (predominantly) or n-type conductivity depending on the nature of its self-doping, making it highly promising for heterojunction design. CuS is often non-stoichiometric, meaning that variations in the oxidation state of its components can result in either excess electrons in its CB or holes in its otherwise filled valence band (VB). This property allows nanosized CuS to be classified as a plasmonic semiconductor due to its self-doping characteristics [33,34].

Plasmonic semiconductors, including CuS, possess extraordinary optoelectronic properties, particularly related to localized surface plasmon resonances (LSPRs) in the near-infrared (NIR) spectral region [35,36]. These characteristics make CuS an excellent candidate for photoelectrochemical (PEC) applications, such as light-assisted water splitting, either as a standalone material [37] or when coupled with titania [38].

Such self-doped particles with controlled morphology were previously prepared through wet synthesis routes with variable degrees of success (see, for instance [39–43]). Achieving the desired non-stoichiometry without contamination from foreign phases required several strategies. Typically, the redox properties of the reaction medium were carefully adjusted to attain the appropriate mixed valence states of copper and/or sulfur while avoiding the formation of impurities such as CuO or metallic Cu. Post-synthesis treatments, such as ion exchange or exposure to a redox atmosphere, were also employed to modify the oxidation states of the copper and sulfur elements [38,40–45].

Given that one of the primary goals of this study was the straightforward and reproducible production of uniformly sized, well-crystallized, and self-doped CuS NPs, we opted for microwave (MW)-assisted polyol synthesis. By varying the MW heating power and time and using either thioacetamide (TAA) or thiourea (ThU) as sulfur sources, we successfully optimized the synthesis to obtain the desired nanostructures. In this method, Cu^{2+} and S^{2-} precursors were dissolved in a polyol solvent, and rapid heating was applied to the reaction medium. This facilitated nucleophilic substitution and condensation reactions while preventing complete reduction, thereby avoiding the contamination of Cu^0 [46], and allowing partial reduction of Cu^{2+} to Cu$^+$ for the desired covellite CuS self-doping.

Additionally, the adsorption of polyol molecules on the surface of the primary particles inhibited their growth and aggregation, enabling effective size control [47]. Finally, using a simple ethanol-based impregnation process, the CuS NPs were deposited onto and between the TiO$_2$ nanofibers (NFs), resulting in the desired CuS-TiO$_2$/Ti architectures.

Structural, optical, and electrochemical characterizations of the resulting CuS-TiO$_2$/Ti composite confirmed the effectiveness of this material and processing approach. These results contribute to the advancement of photoelectrochemical (PEC) technology, highlighting its potential for sustainable hydrogen production.

2. Experiments

2.1. Material Synthesis

TiO$_2$/Ti sheets were prepared by hydrothermal corrosion of commercial Ti plates. In brief, 0.8 × 2 cm Ti metallic plates (Goodfellow, >97%, 1 mm of thickness) were mechanically polished (intermediate polishing) with sandpaper of different granulometries to ensure the removal of the pre-existing oxide layers and eventual contaminants. The samples were washed in ultrasound in water, ethanol, and acetone (10 min each) and air-dried before being submitted to chemical polishing. In practice, the plates were immersed in an oxalic acid aqueous solution (5% w/w, equivalent to 0.6 mol·L^{-1}) and heated at 100 °C for 2 h. They were then washed in water and air dried before their controlled hydrothermal surface oxidation according to already optimized operating conditions [1]: The plates were placed in a Teflon®-lined 120 mL autoclave in an equivolumetric mixture (10 mL total) of H$_2$O$_2$ (30%) and NaOH (10 mol·L^{-1}). The closed autoclave was placed in an oven at 80 °C for 24 h. The plates were rinsed with deionized water, protonated HCL solution (0.1 mol·L^{-1}), and again with deionized water and dried at 80 °C. Finally, a calcination in air took place at a 45 min heating ramp, with a target temperature of 400 °C kept constant for 60 min. Scanning electron microscopy (SEM) confirmed the 1D porous network titania morphology covering the entire titanium sheet (between 0.5 and 1.0 μm in thickness), with ropes of an average diameter of 20–50 nm, interweaving between each other, leading to a highly porous hierarchical structure (Figure S1). Transmission electron microscopy (TEM) evidences the veil-type structure on individual titania NFs, each veil being folded on itself, resulting in a fiber structure with a large specific area (Figure S2) and then a large interaction surface, which is an advantage for the desired catalytic application.

CuS particle synthesis was performed by the MW-assisted polyol process under different conditions. In the presence of sulfide nucleophilic agents, microwave heating allows shortening the reaction, promoting sulfidation instead of total reduction [46]. Two sulfide sources were used (ThU and TAA), and two different operating conditions were explored: low heating power, typically 200 W, for relatively long reaction times (25 to 30 min), and high heating power, namely 1200 W, for very short reaction times (1 to 3 min). In practice, 6.15×10^{-2} mol·L^{-1} of copper (II) acetate were dispersed in 80 mL of ethyleneglycol (EG) with either TAA or ThU (TAA or ThU/copper molar ratio being equal to 1.2). The mixture was vigorously agitated and submitted to intense ultrasound for at least 30 min. The mixture was then transferred to a microwave-adapted reactor and heated in a multiwave Anton Paar microwave oven under constant radiation power. The resulting particles were recovered by centrifugation and washed with ethanol at least three times. They were finally dried at 60 °C overnight in air. The list of the prepared samples is summarized in the supporting information section (Table S1), in which each sample is referenced by adding to CuS the type of sulfur source, microwave power, and heating time. For instance, CuS-ThU-1200-1 corresponds to particles prepared with ThU for a heating time of 1 min under a heating power of 1200 W. All the produced particles are from the covellite structure, as confirmed by Rietveld refinements on all the recorded X-ray diffraction (XRD) patterns (Figure S3 and Table S2). The smallest CuS particles were selected for the final photoanode preparation step to take advantage of their large specific surface area. According to TEM micrographs of ethanolic suspensions containing the variously prepared particles (Figure S4), using a low microwave power of 200 W with a long reaction time (25 min) resulted in larger particle sizes (up to ~40 nm). In contrast, a higher power of 1200 W with a shorter reaction time (1 min) significantly reduced the particle size (down to ~7 nm). Additionally, for a given reaction time, particles synthesized with the thiourea (ThU) precursor were consistently smaller than those produced with thioacetamide (TAA), due to the faster decomposition of ThU in the reaction medium [47]. As a result, the CuS-ThU-1200-1 particles, with a typical size of 7–8 nm, were chosen for the fabrication of the CuS-TiO$_2$/Ti photoanode.

The previously prepared TiO$_2$/Ti substrates were fully immersed in a dilute CuS impregnation solution (3 mg of CuS in 4 mL of ethanol), sonicated for 10 min, and then left to rest overnight. This low concentration was deliberately chosen to allow for a performance comparison between our engineered CuS-TiO$_2$/Ti photoanode and a similarly prepared photoanode in which the less toxic CuS co-catalysts were replaced with the more toxic CdS ones [12,22]. After impregnation, the plates were rinsed with ethanol, dried at 80 °C for 1 h, and stored under standard conditions without requiring any special handling.

2.2. Material Characterization

The structure of all the prepared samples was examined by XRD using two diffractometers (Panalytical, Almelo, Netherlands), an Empyrean equipped with a Cu Kα X-ray source (1.5418 Å) operating in the w-2θ (w = 1°) geometry for the plates and an X'pert Pro equipped with a Co Kα X-ray source (1.7889 Å) operating in the θ-θ geometry for the powders. The collected patterns were analyzed thanks to Highscore+ software version 5.2.0 (PANAYTICAL©, Almelo, The Netherlands).

The chemical composition was investigated by X-ray photoelectron spectroscopy (XPS) on an Escalab250 instrument (Thermo-VG, East Grinstead, UK) equipped with an Al-K$_\alpha$ X-ray source (1486.6 eV). The pass energy was maintained at 200 eV for the survey scan (step size = 1 eV) and at 80 eV for the high-resolution spectra (step size = 0.1 eV). The spectra were calibrated against the (C-C/C-H) C 1s component set at 285 eV, and their analysis was achieved thanks to Avantage software, version 5.9902 (Thermo Scientific™, Boston, MA, USA).

The exact morphology of CuS NPs and TiO_2 NFs was checked by TEM using a JEM 2100 Plus microscope (JEOL, Tokyo, Japan) operating at 200 kV. Additionally, SEM was carried out on the as-produced pristine TiO_2/Ti and composite CuS-TiO_2/Ti photoanodes, using a Gemini SEM 360 microscope (ZEISS, Jena, Germany) operating at 5 kV to check their general morphology. The microscope is also equipped with an Oxford Instrument (Abingdon, UK) energy-dispersive X-ray spectroscopy (EDX) detector (Ultim Max 170 mm^2 detector), allowing chemical analysis, including chemical mapping. All The recorded micrographs were analyzed by ImageJ software version 1.54j (open source).

2.3. Photoelectrochemical Assays

Each prepared photoanode, TiO_2/Ti or CuS-TiO_2/Ti, was employed as a working electrode (WE) in a home-made quartz single-compartment PEC cell (Figure 1), using an Ag/AgCl reference electrode (RE), a Pt wire counter electrode (CE), and a Na_2SO_4 aqueous electrolyte ([SO_4] = 0.5 M, pH = 7). In practice, the I-V curves, thanks to a AUTOLAB PGSTAT12 scanning potentiostat (Metrohm Instrument, Herisau, Switzerland), were collected. Prior to all experiments, the electrolyte was purged by Argon from dissolved dioxygen. To simulate a solar light exposition, a 150 W Xenon lamp (ORIEL instruments, Bozeman, MO, USA) was used, fixing the area of WE illumination to 0.7×1.0 cm^2.

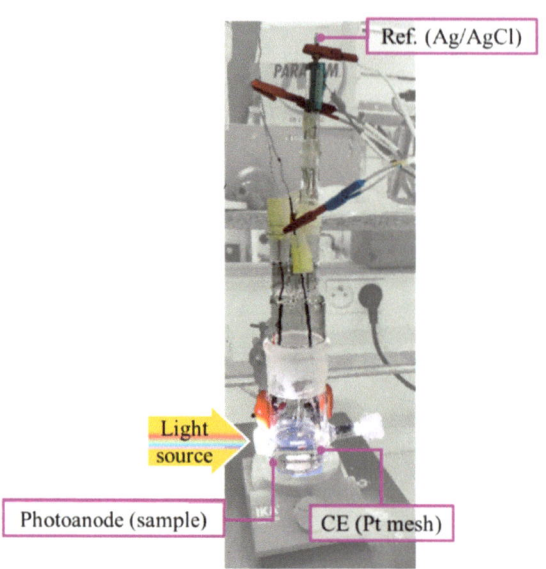

Figure 1. Home-made single-compartment quartz PEC cell, working in a classic 3-electrode configuration, using Ag/AgCl RE and Pt CE.

Prior to PEC measurements, the UV-visible diffuse reflectance spectra of the produced composites were recorded on a Lambda 1050 spectrophotometer (PerkinElmer, Shelton, CT, USA) equipped with a PTFE-coated integration sphere.

3. Results and Discussion

3.1. Photoanode Engineering

The elaboration of the CuS-TiO$_2$/Ti substrates consisted of three main steps involving Ti plate-controlled corrosion to produce a well-adherent thick, porous anatase coating on a conductive substrate, polyol CuS particle synthesis optimization to obtain ultrafine co-catalysts (less than 10 nm in size), and an easy-to-achieve impregnation route, tacking advantage from the abundance of pores and the high surface-to-volume ratio of pristine TiO$_2$/Ti.

The efficiency of the photoanode material processing was first checked by XRD analysis (Figure 2). The recorded pattern of CuS-TiO$_2$/Ti matched very well with that of pure TiO$_2$ and Ti phases. Indeed, all the diffraction peaks were fully indexed in the tetragonal anatase structure (ICDD No. 00-021-1272) and the hexagonal titanium one (ICDD No. 00-044-1294) without clear evidence of CuS signature due to its low content and/or its ultrasmall crystal size.

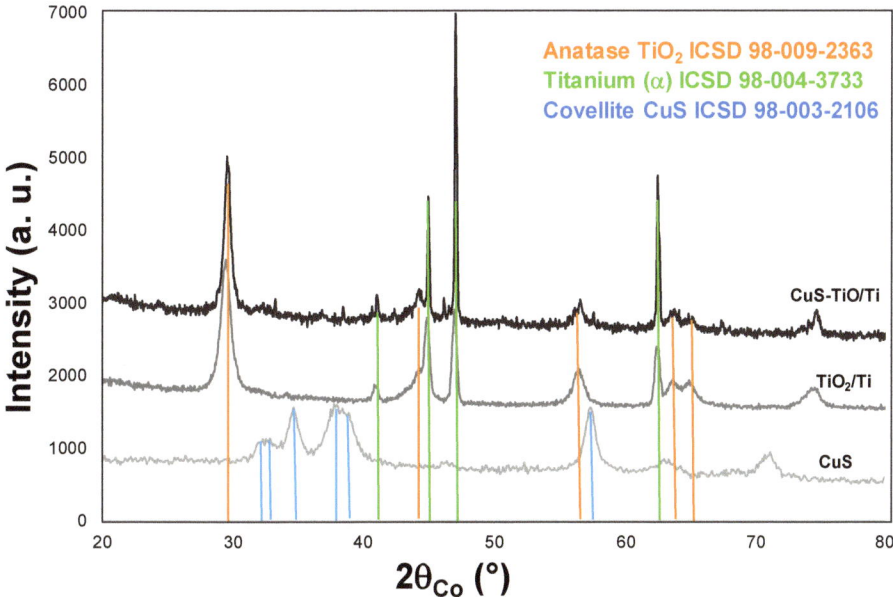

Figure 2. XRD patterns of as-prepared CuS-TiO$_2$/Ti, TiO$_2$/Ti, and CuS. The peak positions of TiO$_2$ (anatase), Ti (α), and CuS (covellite) references are given for information.

To confirm the presence of CuS particles, the engineered photoanode was observed by SEM, and the recorded SEM micrographs were compared to those collected on pristine TiO$_2$/Ti. A simple contrast lecture of the two types of images evidenced some differences on some titania fiber nodes (Figure 3). Focusing on such a zone, EDS chemical mapping confirmed the simultaneous presence of copper and sulfur elements at this area at almost the same concentration (Figure 4).

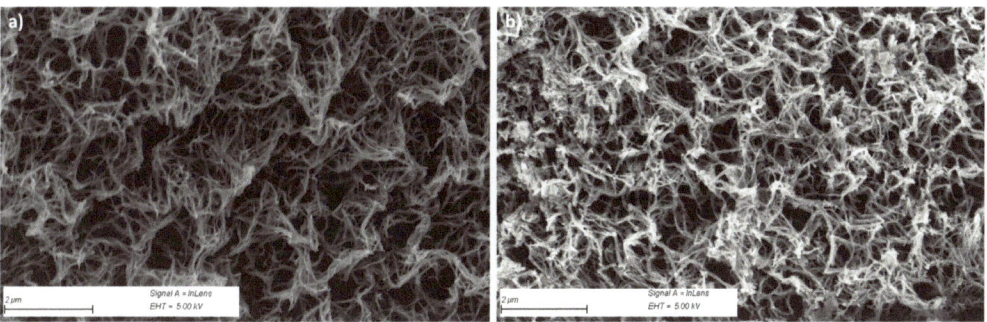

Figure 3. Top view SEM micrographs of TiO$_2$/Ti (**a**) before and (**b**) after CuS impregnation, highlighting the presence of an additional contrast at some TiO$_2$ fiber nodes.

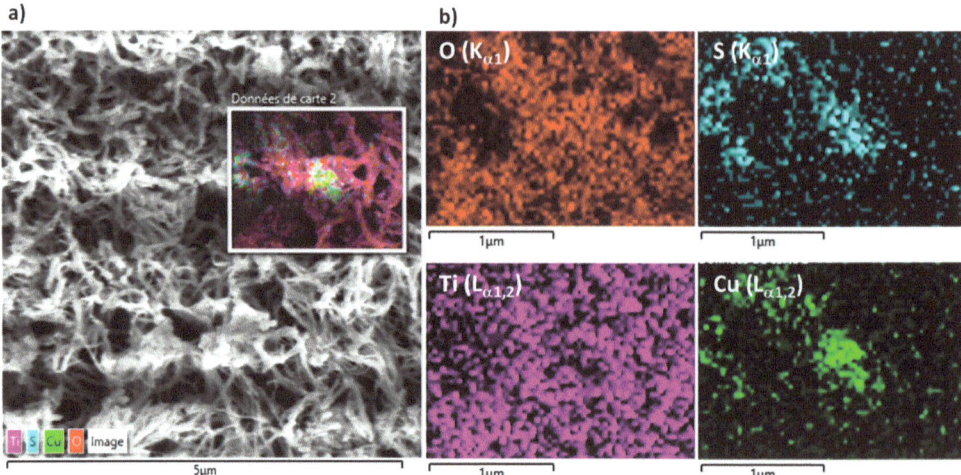

Figure 4. SEM-EDX analysis of the CuS-TiO$_2$/Ti sample: (**a**) Z-contrasting top view SEM micrograph highlighting a TiO$_2$ fiber noddle on which an assembly of CuS particles is aggregated, (**b**) EDS chemical mapping confirming the copper and sulfur element co-concentration in the selected area, in agreement with the presence of CuS particles.

A semi-quantitative EDS analysis of CuS-TiO$_2$/Ti confirmed the presence of copper and sulfur elements at very low but non-zero contents compared to titanium and oxygen elements (Figure S5), leading to a whole Cu/Ti content of about 0.5 at.-%. Such an atomic ratio aligns well with the low concentration of the CuS impregnation solution used in the photoanode preparation. This ratio is also comparable to the Cd/Ti atomic ratio in our previously studied CdS-TiO$_2$/Ti photoanode, making a performance comparison between the two systems in terms of PEC efficiency both relevant and meaningful.

Additionally, a comparison of the Cu/Ti atomic ratio from EDS with that inferred from XPS analysis confirmed that a significant portion of the impregnated CuS particles reside on the outer surface of the TiO$_2$ fibers (15.4 at.% vs. 0.5 at.%), which is advantageous for our intended application. The survey XPS spectrum of CuS-TiO$_2$/Ti, compared to those of pristine TiO$_2$/Ti and CuS (Figure 5a), confirms the presence of all expected elements—Ti and O for the TiO$_2$ phase, and Cu and S for the CuS phase. While there were no notable differences in the respective bonding energies between samples, a significant variation in the Cu2p and S2p peak intensities was observed. Specifically, the surface Cu concentration on CuS-TiO$_2$/Ti was 2.9 at.%, compared to 23.2 at.% on the surface of pristine CuS.

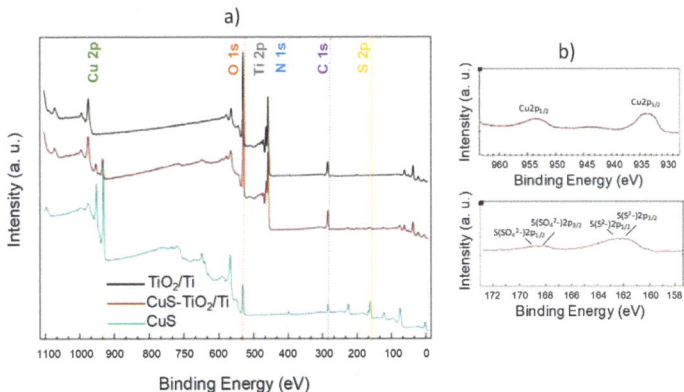

Figure 5. (a) Survey XPS spectra of CuS-TiO$_2$/Ti (brown line), TiO$_2$/Ti (black line), and CuS (blue-green line). (b) Cu 2p and S 1s XPS high-resolution spectra of CuS-TiO$_2$/Ti (brown line).

A focus on the high-resolution Cu 2p signal recorded on CuS-TiO$_2$/Ti (Figure 5b) compared to that of pristine CuS (Figure S6) confirms that copper is at the particle surface divalent with Cu 2p$_{1/2}$ and Cu 2p$_{3/2}$ binding energies of 952.5 and 932.4 eV, respectively, close to the values reported in the literature for CuS [48,49] and Cu$_2$S [50,51] phases. CuS also exhibits a small shake-up or multiplet splitting structure, while Cu$_2$S does not [48–52]. This feature agrees with the formation of CuS without excluding the presence of Cu$^+$ species. Additionally, the Cu LMM peaks (Figure S7) recorded on all the prepared CuS particles, including those used for the preparation of CuS-TiO$_2$/Ti, exhibit a peak shape completely different from that usually observed on Cu^0 species [52], confirming the absence of copper metal. Moreover, the slight non-stoichiometry measured by XPS on the CuS particles before and after their attachment by impregnation to the pristine TiO$_2$/Ti (Table 1) agrees fairly with self-doping, which may result from a partial substitution of Cu^{2+} cations by monovalent Cu$^+$ ones within the covellite lattice.

Table 1. Recapitulative table of binding energies and atomic compositions for CuS-TiO$_2$/Ti photoanodes and their pristine TiO$_2$/Ti and CuS counterparts.

	Binding Energy (eV)			Content (at.- %)		
	TiO$_2$/Ti	CuS-TiO$_2$/Ti	CuS	TiO$_2$/Ti	CuS-TiO$_2$/Ti	CuS
C 1s (C-C/C-H)	284.8	284.8	284.8	17.6	19.0	13.4
C 1s (C-O)	286.4	286.5	286.5	3.9	4.9	5.9
C 1s (C=O)	288.8	288.5	289.1	2.0	2.0	2.0
Cu LMM	-	565.3	568.9	-	-	-
Cu 2p	-	933.6	932.2	-	2.9	23.2
N 1s	400.1	399.7	399.8	0.4	0.6	4.2
O 1s	529.8	530.0	532.2	53.7	49.5	23.7
S^{2-} 2p	-	162.2	162.5	-	1.7	21.6
SO$_4^{2-}$ 2p	-	168.5	168.9	-	0.4	5.8
Ti 2p	458.5	458.6	-	22.4	19.0	-

Assuming that all these features are representative of the whole volume of all CuS particles (the 7–8 nm average size of CuS particles is smaller than the 10–12 nm XPS analysis depth), one may conclude in favor of their self-doping, giving then of the properties of plasmonic semiconductors.

Also, the S2p high-resolution XPS spectra of both CuS (Figure S6) and CuS-TiO$_2$/Ti (Figure 5b) are quite similar. Their total intensities are of course different, but both exhibit a doublet at 163.5 (2p$_{1/2}$) and 162.3 eV (2p$_{3/2}$) characteristics of sulfide S^{2-} species in

CuS [48,49] or Cu_2S [50,51] phases. Interestingly, both exhibit a supplementary contribution: a broad and small in intensity peak at 168.6 eV usually attributed to sulfate SO_4^{2-} anions, suggesting a weak surface oxidation with the production of a thin $CuSO_4$ passivation layer. In other words, the composition of the analyzed copper sulfide particles is consistent with a $CuS@CuSO_4$ core-shell nanostructure. Comparing the intensity of the S^{2-} $2p_{3/2}$ and SO_4^{2-} $2p_{3/2}$ XPS peaks allows us to estimate that, approximatively, the fourth of the involved sulfur atoms are in the form of sulfate, in agreement with a very thin protective copper sulfate layer.

If the former XPS analysis confirmed the presence of CuS particles on the surface of the $CuS-TiO_2/Ti$ sample, it also suggested that the chosen impregnation route did not affect the chemical state of titanium cations on the surface of the titania coating. Indeed, there are no significant differences between the Ti 2p XPS profiles of TiO_2/Ti and $CuS-TiO_2/Ti$, as well as between their O 1s XPS profiles (Figure S6), agreeing very well with the TiO_2 oxide nature of the outer layer of the titanium plates [1,5,53].

Finally, the optical absorption spectrum of the engineered $CuS-TiO_2/Ti$ photoanode was measured in diffuse reflectance and compared to that of pristine TiO_2/Ti, recorded in diffuse reflectance as well, and that of pristine CuS, recorded in a transmission scheme (Figure 6). Regarding the semiconducting nature of TiO_2/Ti, the typical anatase band-to-band signature at around 300 nm was identified, with a band-gap value inferred from Tauc plots of about 3.2 eV in pristine TiO_2/Ti and 2.6 eV in $CuS-TiO_2/Ti$. The last small value is not at all the consequence of a gap decrease but is the consequence of a more complex composite band diagram. Indeed, the $CuS-TiO_2/Ti$ spectrum is the combination of those of pristine TiO_2/Ti and CuS, with absorption capabilities ranging from UV to NIR spectral ranges, due to the photo-excitation of both titania and copper sulfite semiconductors. The anatase and the covellite band-to-band absorptions (around 300 [1,5] and 400 nm [30], respectively) superposed to the self-doped CuS LSPR absorption (around 1200 nm [31,32]) explain together the optical properties of our engineered photoanode. Clearly, the amount of CuS particles deposited on TiO_2/Ti by impregnation, even small, appeared large enough to induce a widened light absorption, which is fruitful for improved PEC responses.

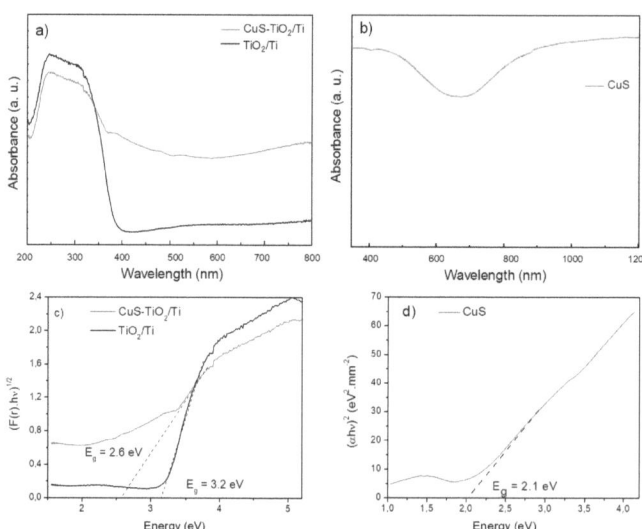

Figure 6. UV-Vis-NIR absorption spectra of (**a**) $CuS-TiO_2/Ti$ and TiO_2/Ti recorded in total reflectance mode compared to (**b**) that of pristine CuS recorded in transmission. (**c,d**) The Tauc plots inferred from the previous data are given for band-gap determination. The lamp change from UV to visible range during spectra acquisition proceeded at 320 nm.

3.2. Photoanode PEC Properties

Hydrogen photo-generation activity of the as-synthetized CuS-TiO$_2$/Ti photoanode and its TiO$_2$/Ti parent was carried out under a Xenon lamp irradiation using a passive and neutral electrolyte. Interestingly, operating in a passive electrolyte, an intermittent illumination of CuS-TiO$_2$/Ti provides a higher photocurrent than pristine TiO$_2$/Ti, whatever the applied bias (Figure 7). The chronoamperometry under intermittent lighting (black arrows indicating the beginning of dark periods and the orange arrows indicating the beginning of illuminated periods) shows that the sensible increase in the photogenerated current is stable at 1.23 V (vs. Ag/AgCl), and while there is a decrease in the photocurrent during the first minute of illuminated periods, the original values are restored after a dark period.

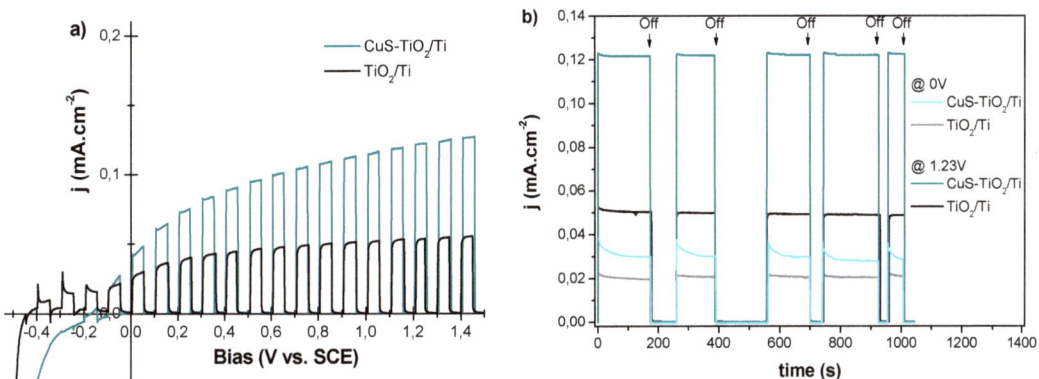

Figure 7. (a) Linear sweep voltammetry (10 mV.s^{-1}) and (b) chronoamperometry of TiO$_2$/Ti (black line) are CuS-TiO$_2$/Ti (blue-green line) in a passive Na$_2$SO$_4$ (0.5 M) electrolyte.

By comparing the behavior of the bare and impregnated photoanodes under 0 V and 1.23 V bias, we can infer that the improved photocurrent is due to a higher amount of photogenerated charge carriers, favored by the increase of absorbed photons, promoted by CuS NPs. CuS-TiO$_2$/Ti absorbs more light, in a wider spectral range, than pristine TiO$_2$/Ti, creating a sufficient number of electron−hole pairs. The electrons can then be transferred from CuS CB to that of TiO$_2$ as summarized hereafter:

$$CuS + h\nu \rightarrow CuS(e^- + h^+) \quad (1)$$

$$e^-(CB_{CuS}) + TiO_2 \rightarrow TiO_2(e^-) \quad (2)$$

$$h^+(VB_{TiO_2}) + H_2O \rightarrow 2H^+ + \frac{1}{2}O_2 \quad (3)$$

$$h^+(VB_{TiO_2}) + CuS \rightarrow CuS(h^+) \quad (4)$$

$$h^+(VB_{CuS}) + H_2O \rightarrow 2H^+ + \frac{1}{2}O_2 \quad (5)$$

The collected electrons in TiO$_2$ CB were then transferred through the external circuit to the Pt cathode to achieve the reduction of aqueous protons into hydrogen gas (Figure 8).

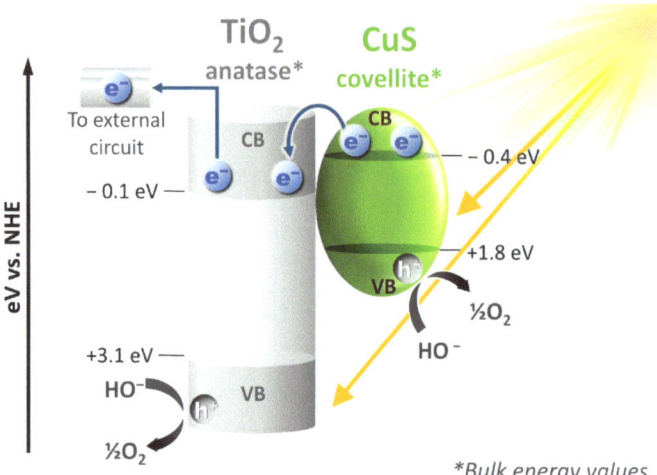

Figure 8. General scheme of the energy band diagram of bulk TiO_2 anatase and CuS covellite versus the normal hydrogen electrode (NHE), highlighting the reaction of their VB holes with water molecules to produce O_2 and the collection of their CB electrons for their transfer to the external circuit in a standard PEC cell. To build this diagram, band gap energies and band positions versus NHE of anatase TiO_2 and covellite CuS were inferred from [54,55], respectively.

These findings align with results from a few research groups studying CuS-TiO_2 systems, which remain relatively underexplored in the literature on photocatalytic hydrogen generation compared to metal chalcogenide-based titania nanocomposites, such as the CdS-TiO_2 system. This is despite the well-documented acute toxicity of (see, for instance, [19–21] and the references therein).

To the best of our knowledge, notable results have been reported by Chandra et al. [56], who prepared their composites by a hydrothermal and a solution-based process. Operating by photocatalysis (PC) in a sacrificial Na_2S (0.25 M)-Na_2SO_3 (0.25 M) electrolyte, they succeeded in producing 1262 µmol of H_2 per hour and per gram of catalyst, more than 10 and 9 times higher than that by pristine TiO_2 and pristine CuS powders under Xe lamp irradiation, respectively. There are also results reported by Jia et al. [57], who decorated TiO_2 nanowire arrays grown on conductive substrate by CuS nanoclusters by successive ionic layer adsorption and reaction (SILAR method). Operating by photo-electrocatalysis (PEC) in a passive Na_2SO_4 (1.00 M) electrolyte, they demonstrated an increased light absorption and an efficient charge separation leading to an improved photocurrent. They succeeded in obtaining within the same setup a photocurrent density 5 times higher than that of pristine TiO_2 at a bias of 0.35 V. One may also cite the results of Liu et al. [58], who successfully constructed a CuS/TiO_2 heterojunction using metal-organic framework (MOF)-derived TiO_2 as a substrate. They pointed out that CuS/TiO_2 exhibited excellent bifunctional PC activity without noble metal cocatalysts. They typically evidenced H_2 production and benzylamine oxidation in a coupled experiment, with a H_2 evolution activity of the CuS/TiO_2 17.1 and 29.5 times higher than that of TiO_2 and CuS, respectively. Wang et al. [59] also investigated CuS/TiO_2 photocatalysts, prepared via a high-temperature hydrothermal method, and evaluated their photocatalytic activity. They demonstrated that loading TiO_2 with 1 wt.-% CuS significantly enhanced its photocatalytic performance for water decomposition to hydrogen in a methanol aqueous solution under Xe lamp irradiation. The CuS/TiO_2 photocatalysts produced approximately 570 µmol of H_2 per hour, which is 32 times higher than that produced by pristine TiO_2.

Clearly, in all these studies and in others (Table S3), widened light absorption and an efficient charge separation were systematically reported. All converged, placing CuS as one of the most interesting metal chalcogenide titania co-catalysts for water splitting.

To support these scientific advances, we compared the performance of the CuS-TiO$_2$/Ti photoanode with that of a similarly prepared CdS-TiO$_2$/Ti photoanode [12,22]. The main difference between the two is the size of the particles, with CdS having a particle size of 3 nm. Both photoanodes were tested using the same photoelectrochemical (PEC) setup. Interestingly, the photocurrent measured for the CuS-TiO$_2$/Ti photoanode was consistently higher than that for the CdS-TiO$_2$/Ti photoanode. This is attributed to the broader light absorption range of CuS (Figure 9), indicating that CuS is a more effective co-catalyst compared to CdS.

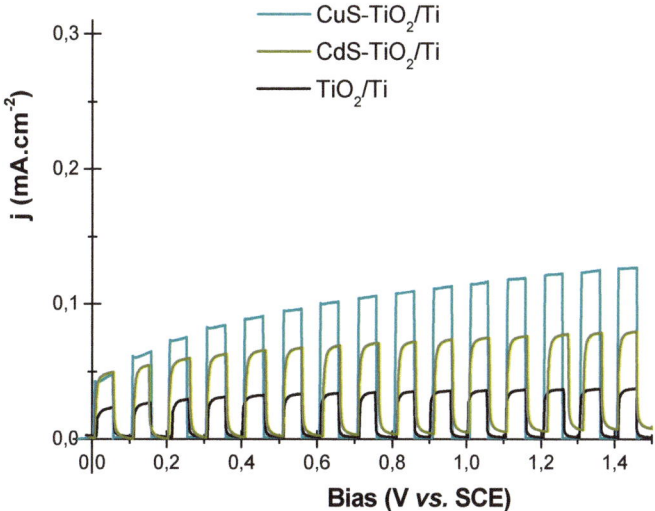

Figure 9. Linear sweep voltammetry (10 mV.s^{-1}) of CuS-TiO$_2$/Ti (blue-green line), CdS-TiO$_2$/Ti (green line), and TiO$_2$/Ti (black line) in a passive Na$_2$SO$_4$ (0.5 M) electrolyte, focusing on the 0 to 1.5 V bias range.

4. Conclusions

In conclusion, the integration of CuS nanoparticles (NPs) with TiO$_2$ nanofibers (NFs) has proven to be a promising approach for achieving efficient photoelectrochemical (PEC) responses in water splitting and hydrogen generation, in good alignment with the relevant literature. The synergistic properties of these nanomaterials enable excellent light absorption, effective charge separation, and efficient electron transport, leading to significant improvements in PEC performance.

By optimizing the microwave-assisted polyol process conditions, self-doped covellite CuS particles with sizes of 7–8 nm, which absorb in the visible and near-infrared (NIR) spectral ranges, were successfully produced without foreign contaminants. These particles were effectively integrated with TiO$_2$ NFs supported on a titanium substrate, which was prepared through controlled metal plate corrosion (hydrothermal treatment followed by calcination). The simple ethanol-based impregnation method proved sufficient for creating the CuS-TiO$_2$/Ti semiconductive hetero-nanostructures.

A CuS concentration as low as 0.5 at.-% in the composite was sufficient to achieve photocurrents of 0.030 and 0.122 mA/cm^2 at 0 V and 1.23 V, respectively. In comparison, the photocurrents measured for pristine TiO$_2$/Ti under the same PEC conditions were 0.020 and 0.051 mA/cm^2. Notably, the photocurrent with the CuS co-catalyst was comparable to that obtained with toxic CdS at 0 V and significantly higher at 1.23 V (0.122 vs.

0.082 mA/cm^2). These results highlight that the selected materials and the employed synthetic approaches offer a novel and effective pathway for developing sustainable hydrogen production systems.

Supplementary Materials: The following supporting information can be downloaded at: https://www.mdpi.com/article/10.3390/nano14191581/s1, Figure S1: SEM (a) top view and (b) cross view of the as-prepared TiO$_2$/Ti sheet confirming the 1D morphology of the formed titania, covering all the Ti substrate surface; Figure S2: TEM micrographs recorded on an (a) assembly and an (b) individual representative TiO$_2$ NF separated by sonication from the as-produced TiO$_2$/Ti sheets in an ethanolic solution. Titania fibers appear as veils folded on themselves; Figure S3: (a) XRD patterns of CuS particles prepared using TAA (up) and ThU (bottom) reagents while applying a microwave heating power of 200 and 1200 W for a total reaction time of 25 and 3 min, respectively. (b) Results of Rietveld refinements (using MAUD software) performed on the XRD pattern of CuS-ThU-1200-1 particles to illustrate the quality of the fits: the experimental pattern (black scatter) and the calculated one (green line) are perfectly superposed with a residue curve, defined as the difference between the experimental and calculated diffractograms, close to zero (blue line). The inferred crystallite shape is also given for information [60]; Figure S4: SEM images recorded on (a) CuS-TAA-200-25 and (b) CuS-ThU-200-25 particles. TEM images of (c) CuS-TAA61200-3 and (d) CuS-1200-ThU-1200-3 particles. (e) TEM micrograph of CuS-ThU-1200-1 particles and (f) HRTEM image of some representative CuS-ThU-1200-1 particles; Figure S5: SEM-EDS analysis of CuS-TiO$_2$/Ti, focusing on TiO$_2$ fiber nodes, where CuS particles seem to accumulate, leading to an average Cu/Ti atomic ratio of 0.5 at.-%; Figure S6. Ti 2p and O 1s high-resolution XPS spectra recorded on pristine TiO$_2$/Ti (black) and Cu2P and S 2p high-resolution XPS spectra recorded on pristine CuS (blue-green line); Figure S6: Ti 2p and O 1s high-resolution XPS spectra recorded on pristine TiO$_2$/Ti (black) and Cu2P and S 2p high-resolution XPS spectra recorded on pristine CuS (blue-green line); Figure S7: S 2p, Cu 2p, and Cu LMM high-resolution XPS spectra recorded on CuS NPs produced using ThU or TAA sulfur source for a microwave heating power of 1200 W along 1 min of reaction time; Table S1: List of prepared CuS samples and their main synthesis, MW-assisted polyol synthesis; Table S2: Main Rietveld refined structural parameters and their related reliability fit factors. The overall fit quality is described by a weighted profile (R_{wp}), expected profile (R_{exp}), and Bragg R-value (R_B) close to 1; Table S3: Comparison of the PEC performances of our engineered photoanode with those of CuS-TiO$_2$-based literature [38,56,57,61–67].

Author Contributions: L.C.: Conceptualization, Data curation, Formal analysis, Investigation, Methodology, Visualization, Writing—review & editing; S.C.: Conceptualization, Data curation, Formal analysis, Writing—review & editing; B.D.: Data curation, Formal analysis (he prepared and characterized CuS partucles); S.G.-D.: Data curation and Formal analysis (she performed SEM-EDS measurements and analyzed the collected data and images); S.N.: Data curation, Formal analysis (she performed XRD experiments and analyzed the collected data); F.M.: Conceptualization, Supervision, Visualization, Writing—review & editing; S.A.: Conceptualization, Supervision, Methodology, Visualization, Validation, Writing—review & editing, Funding acquisition, Project administration, Writing—original draft. All authors have read and agreed to the published version of the manuscript.

Funding: ANR (Agence Nationale de la Recherche) and CGI (Commissariat à l'Investissement d'Avenir) are gratefully acknowledged for financial support of this work through Labex SEAM (Science and Engineering for Advanced Materials and devices), ANR-10-LABX-096, and ANR-18-IDEX-0001 grants.

Data Availability Statement: The raw data supporting the conclusions of this article will be made available by the authors on request.

Acknowledgments: The authors would like to acknowledge Guillaume Thoraval and Tom Chevry (Université Paris Cité), who crafted the glassware necessary for the PEC experiments. They are also grateful to Patricia Beaunier (Sorbonne Université), who managed all TEM experiments.

Conflicts of Interest: The authors declare no conflict of interest.

References

1. Chaguetmi, S.; Achour, S.; Mouton, L.; Decorse, P.; Nowak, S.; Costentin, C.; Mammeri, F.; Ammar, S. TiO_2 nanofibers supported on Ti sheets prepared by hydrothermal corrosion: Effect of the microstructure on their photochemical and photoelectrochemical properties. *RSC Adv.* **2015**, *5*, 95038–95046. [CrossRef]
2. Lianos, P. Review of recent trends in photoelectocatalytic conversion of solar energy to electricity and hydrogen. *Appl. Catal. B Environ.* **2017**, *210*, 235–254. [CrossRef]
3. Egerton, T.A. Does photoelectrocatalysis by TiO_2 work? *J. Chem. Technol. Biotechnol.* **2011**, *86*, 1024–1031. [CrossRef]
4. Tsao, C.-W.; Fang, M.-J.; Hsu, Y.-J. Modulation of interfacial charge dynamics of semiconductor heterostructures for advanced photocatalytic applications. *Coord. Chem. Rev.* **2021**, *438*, 213876. [CrossRef]
5. Lettieri, S.; Pavone, M.; Fioravanti, A.; Amato, L.S.; Maddalena, P. Charge Carrier Processes and Optical Properties in TiO_2 and TiO_2-Based Heterojunction Photocatalysts: A Review. *Materials* **2021**, *14*, 1645. [CrossRef] [PubMed]
6. Marcus, R.A. On the theory of electron-transfer reactions. VI. Unified treatment for homogeneous and electrode reactions. *J. Chem. Phys.* **1965**, *43*, 679–701. [CrossRef]
7. Gao, Y.Q.; Marcus, R.A. On the theory of electron transfer reactions at semiconductor/liquid interfaces. II. A free electron model. *J. Chem. Phys.* **2000**, *113*, 6351–6360. [CrossRef]
8. Tvrdy, K.; Frantsuzov, P.A.; Kamat, P.V. Photoinduced electron transfer from semiconductor quantum dots to metal oxide nanoparticles. *Proc. Natl. Acad. Sci. USA* **2010**, *108*, 29–34. [CrossRef]
9. Moridon, S.N.F.; Arifin, K.; Yunus, R.M.; Minggu, L.J.; Kassim, M.B. Photocatalytic water splitting performance of TiO_2 sensitized by metal chalcogenides: A review. *Ceram. Int.* **2022**, *48*, 5892–5907. [CrossRef]
10. Mamiyev, Z.; Balayeva, N.O. Metal Sulfide Photocatalysts for Hydrogen Generation: A Review of Recent Advances. *Catalysts* **2022**, *12*, 1316. [CrossRef]
11. Keimer, B.; Moore, J.E. The physics of quantum materials. *Nat. Phys.* **2017**, *13*, 1045–1055. [CrossRef]
12. Chaguetmi, S.; Mammeri, F.; Pasut, M.; Nowak, S.; Lecoq, H.; Decorse, P.; Costentin, C.; Achour, S.; Ammar, S. Synergetic effect of CdS quantum dots and TiO_2 nanofibers for photoelectrochemical hydrogen generation. *J. Nanopart. Res.* **2013**, *15*, 2140. [CrossRef]
13. Khatamian, M.; Oskoui, M.S.; Haghighi, M.; Darbandi, M. Visible-light response photocatalytic water splitting over CdS/TiO_2 and $CdS-TiO_2$/metalosilicate composites. *Int. J. Energy Res.* **2014**, *38*, 1712–1726. [CrossRef]
14. Manchwari, S.; Khatter, J.; Chauhan, R. Enhanced photocatalytic efficiency of TiO_2/CdS nanocomposites by manipulating CdS suspension on TiO_2 nanoparticles. *Inorg. Chem. Commun.* **2022**, *146*, 110082. [CrossRef]
15. Nguyen, V.N.; Doan, M.T.; Nguyen, M.V. Photoelectrochemical water splitting properties of CdS/TiO_2 nanofibers-based photoanode. *J. Mater. Sci. Mater. Electron.* **2019**, *30*, 926–932. [CrossRef]
16. Shen, J.; Meng, Y.; Xin, G. CdS/TiO_2 nanotubes hybrid as visible light driven photocatalyst for water splitting. *Rare Met.* **2011**, *30*, 280–283. [CrossRef]
17. Li, C.; Yuan, J.; Han, B.; Jiang, L.; Shangguan, W. TiO_2 nanotubes incorporated with CdS for photocatalytic hydrogen production from splitting water under visible light irradiation. *Int. J. Hydrogen Energy* **2010**, *35*, 7073–7079. [CrossRef]
18. Sharma, K.; Hasija, V.; Malhotra, M.; Verma, P.K.; Khan, A.A.P.; Thakur, S.; Van Le, Q.; Quang, H.H.P.; Nguyen, V.-H.; Singh, P.; et al. A review of CdS-based S-scheme for photocatalytic water splitting: Synthetic strategy and identification techniques. *Int. J. Hydrogen Energy* **2024**, *52*, 804–818. [CrossRef]
19. Varmazyari, A.; Taghizadehghalehjoughi, A.; Sevim, C.; Baris, O.; Eser, G.; Yildirim, S.; Hacimuftuoglu, A.; Buha, A.; Wallace, D.R.; Tsatsakis, A.; et al. Cadmium sulfide-induced toxicity in the cortex and cerebellum: In vitro and in vivo studies. *Toxicol. Rep.* **2020**, *7*, 637–648. [CrossRef]
20. Hossain, S.T.; Mukherjee, S.K. Toxicity of cadmium sulfide (CdS) nanoparticles against Escherichia coli and HeLa cells. *J. Hazard. Mater.* **2013**, *260*, 1073–1082. [CrossRef] [PubMed]
21. Mahan, N.K.; Hassan, M.A.M.; Mohammed, A.H.; Khalee, R.I. Neuronal Toxicity of CdS Nanoparticles Prepared by Laser Ablation and their Effect on Liver. *Mater. Sci. Forum* **2021**, *1039*, 537–556. [CrossRef]
22. Chaguetmi, S.; Chaperman, L.; Nowak, S.; Schaming, D.; Lau-Truong, S.; Decorse, P.; Beaunier, P.; Costentin, C.; Mammeri, F.; Achour, S.; et al. Photoelectrochemical properties of ZnS- and $CdS-TiO_2$ nanostructured photocatalysts: Aqueous sulfidation as a smart route to improve catalyst stability. *J. Photochem. Photobiol. A Chem.* **2018**, *356*, 489–501. [CrossRef]
23. Liu, Q.; Liu, Y.; Lei, T.; Tan, Y.; Wu, H.; Li, J. Preparation and characterization of nanostructured titanate bioceramic coating by anodization–hydrothermal method. *Appl. Surf. Sci.* **2015**, *328*, 279–286. [CrossRef]
24. Wang, H.; Lai, Y.-K.; Zheng, R.-Y.; Bian, Y.; Lin, C.-J.; Zhang, K.-Q. Tuning the surface microstructure of titanate coatings on titanium implants for enhancing bioactivity of implants. *Int. J. Nanomed.* **2015**, *10*, 3887–3896. [CrossRef] [PubMed]
25. Wu, Y.; Long, M.; Cai, W.; Dai, S.; Chen, C.; Wu, D.; Bai, J. Preparation of photocatalytic anatase nanowire films by *in situ* oxidation of titanium plate. *Nanotechnology* **2009**, *20*, 185703–185711. [CrossRef] [PubMed]
26. Feng, T.; Yam, F. The influence of hydrothermal treatment on TiO_2 nanostructure films transformed from titanates and their photoelectrochemical water splitting properties. *Surf. Interfaces* **2023**, *38*, 102767. [CrossRef]
27. Xu, Y.; Zhang, M.; Zhang, M.; Lv, J.; Jiang, X.; He, G.; Song, X.; Sun, Z. Controllable hydrothermal synthesis, optical and photocatalytic properties of TiO_2 nanostructures. *Appl. Surf. Sci.* **2014**, *315*, 299–306. [CrossRef]
28. van Dillen, A.; Terörde, R.J.; Lensveld, D.J.; Geus, J.W.; de Jong, K.P. Synthesis of supported catalysts by impregnation and drying using aqueous chelated metal complexes. *J. Catal.* **2003**, *216*, 257–264. [CrossRef]

29. Campelo, J.M.; Luna, D.; Luque, R.; Marinas, J.M.; Romero, A.A. Sustainable Preparation of Supported Metal Nanoparticles and Their Applications in Catalysis. *ChemSusChem* **2009**, *2*, 18–45. [CrossRef]
30. Jang, J.S.; Kim, H.G.; Lee, J.S. Heterojunction semiconductors: A strategy to develop efficient photocatalytic materials for visible light water splitting. *Catal. Today* **2012**, *185*, 270–277. [CrossRef]
31. Isik, M.; Terlemezoglu, M.; Gasanly, N.; Parlak, M. Structural, morphological and temperature-tuned bandgap characteristics of CuS nano-flake thin films. *Phys. E Low-Dimens. Syst. Nanostruct.* **2022**, *144*, 115407. [CrossRef]
32. Mohammed, K.A.; Ahmed, S.M.; Mohammed, R.Y. Investigation of structure, optical, and electrical properties of CuS thin Films by CBD technique. *Crystals* **2020**, *10*, 684. [CrossRef]
33. Kwon, Y.-T.; Lim, G.-D.; Kim, S.; Ryu, S.H.; Lim, H.-R.; Choa, Y.-H. Effect of localized surface plasmon resonance on dispersion stability of copper sulfide nanoparticles. *Appl. Surf. Sci.* **2019**, *477*, 204–210. [CrossRef]
34. Nishi, H.; Asami, K.; Tatsuma, T. CuS nanoplates for LSPR sensing in the second biological optical window. *Opt. Mater. Express* **2016**, *6*, 1043–1048. [CrossRef]
35. Agrawal, A.; Cho, S.H.; Zandi, O.; Ghosh, S.; Johns, R.W.; Milliron, D.J. Localized Surface Plasmon Resonance in Semiconductor Nanocrystals. *Chem. Rev.* **2018**, *118*, 3121–3207. [CrossRef] [PubMed]
36. Comin, A.; Manna, L. New materials for tunable plasmonic colloidal nanocrystals. *Chem. Soc. Rev.* **2014**, *43*, 3957–3975. [CrossRef] [PubMed]
37. Fu, W.; Liu, M.; Xue, F.; Wang, X.; Diao, Z.; Guo, L. Facile polyol synthesis of CuS nanocrystals with a hierarchical nanoplate structure and their application for electrocatalysis and photocatalysis. *RSC Adv.* **2016**, *6*, 80361–80367. [CrossRef]
38. El-Gendy, R.A.; El-Bery, H.M.; Farrag, M.; Fouad, D.M. Metal chalcogenides (CuS or MoS_2)-modified TiO_2 as highly efficient bifunctional photocatalyst nanocomposites for green H_2 generation and dye degradation. *Sci. Rep.* **2023**, *13*, 7994. [CrossRef] [PubMed]
39. Ye, H.; Tang, A.; Hou, Y.; Yang, C.; Teng, F. Tunable near-infrared localized surface plasmon resonances of heterostructured $Cu_{1.94}$S-ZnS nanocrystals. *Opt. Mater. Express* **2014**, *4*, 220. [CrossRef]
40. Ding, Y.; Lin, R.; Xiong, S.; Zhu, Y.; Yu, M.; Duan, X. The Effect of Copper Sulfide Stoichiometric Coefficient and Morphology on Electrochemical Performance. *Molecules* **2023**, *28*, 2487. [CrossRef]
41. Heydari, H.; Moosavifard, S.E.; Shahraki, M.; Elyasi, S. Facile synthesis of nanoporous CuS nanospheres for high-performance supercapacitor electrodes. *J. Energy Chem.* **2017**, *26*, 762–767. [CrossRef]
42. Naveed, M.; Younas, W.; Zhu, Y.; Rafai, S.; Zhao, Q.; Tahir, M.; Mushtaq, N.; Cao, C. Template free and facile microwave-assisted synthesis method to prepare mesoporous copper sulfide nanosheets for high-performance hybrid supercapacitor. *Electrochim. Acta* **2019**, *319*, 49–60. [CrossRef]
43. Savarimuthu, I.; Susairaj, M.J.A.M. CuS Nanoparticles Trigger Sulfite for Fast Degradation of Organic Dyes under Dark Conditions. *ACS Omega* **2022**, *7*, 4140–4149. [CrossRef]
44. Ahmad, N.; Alshehri, A.; Ahmad, I.; Shkir, M.; Hasan, P.; Melaibari, A.A. In doping effect on the structural, morphological, optical and enhanced antimicrobial activity of facilely synthesized novel CuS nanostructures. *Surf. Interfaces* **2021**, *27*, 101536. [CrossRef]
45. Rudra, S.; Bhar, M.; Mukherjee, P. Post-synthetic modification of semiconductor nanoparticles can generate lanthanide luminophores and modulate the electronic properties of preformed nanoparticles. *4open* **2023**, *6*, 8. [CrossRef]
46. Fiévet, F.; Ammar-Merah, S.; Brayner, R.; Chau, F.; Giraud, M.; Mammeri, F.; Peron, J.; Piquemal, J.-Y.; Sicard, L.; Viau, G. The polyol process: A unique method for easy access to metal nanoparticles with tailored sizes, shapes and compositions. *Chem. Soc. Rev.* **2018**, *47*, 5187–5233. [CrossRef] [PubMed]
47. Longo, A.V.; Notebaert, B.; Gaceur, M.; Patriarche, G.; Sciortino, A.; Cannas, M.; Messina, F.; von Bardeleben, H.J.; Battaglini, N.; Ammar, S. Photo-Activated Phosphorescence of Ultrafine ZnS:Mn Quantum Dots: On the Lattice Strain Contribution. *J. Phys. Chem. C* **2022**, *126*, 1531–1541. [CrossRef]
48. Hussain, S.; Patil, S.A.; Memon, A.A.; Vikraman, D.; Naqvi, B.A.; Jeong, S.H.; Kim, H.-S.; Kim, H.-S.; Jung, J. CuS/WS_2 and CuS/MoS_2 heterostructures for high performance counter electrodes in dye-sensitized solar cells. *Sol. Energy* **2018**, *171*, 122–129. [CrossRef]
49. Karikalan, N.; Karthik, R.; Chen, S.-M.; Karuppiah, C.; Elangovan, A. Sonochemical Synthesis of Sulfur Doped Reduced Graphene Oxide Supported CuS Nanoparticles for the Non-Enzymatic Glucose Sensor Applications. *Sci. Rep.* **2017**, *7*, 2494. [CrossRef]
50. Biesinger, M.C.; Hart, B.R.; Polack, R.; Kobe, B.A.; Smart, R.S. Analysis of mineral surface chemistry in flotation separation using imaging XPS. *Miner. Eng.* **2007**, *20*, 152–162. [CrossRef]
51. Tang, A.; Qu, S.; Li, K.; Hou, Y.; Teng, F.; Cao, J.; Wang, Y.; Wang, Z. One-pot synthesis and self-assembly of colloidal copper(I) sulfide nanocrystals. *Nanotechnology* **2010**, *21*, 285602. [CrossRef] [PubMed]
52. Biesinger, M.C. Advanced Analysis of Copper X-ray Photoelectron (XPS) Spectra. *Surf. Interface Anal.* **2017**, *49*, 1325–1334. [CrossRef]
53. Zhu, L.; Lu, Q.; Lv, L.; Wang, Y.; Hu, Y.; Deng, Z.; Lou, Z.; Hou, Y.; Teng, F. Ligand-free rutile and anatase TiO_2 nanocrystals as electron extraction layers for high performance inverted polymer solar cells. *RSC Adv.* **2017**, *7*, 20084–20092. [CrossRef]
54. Marschall, R. Semiconductor composites: Strategies for enhancing charge carrier separation to improve photocatalytic activity. *Adv. Funct. Mater.* **2014**, *24*, 2421–2440. [CrossRef]

55. Huerta-Flores, A.M.; Torres-Martínez, L.M.; Moctezuma, E.; Singh, A.P.; Wickman, B. Green synthesis of earth-abundant metal sulfides (FeS_2, CuS, and NiS_2) and their use as visible-light active photocatalysts for H_2 generation and dye removal. *J. Mater. Sci. Mater. Electron.* **2018**, *29*, 11613–11626. [CrossRef]
56. Chandra, M.; Bhunia, K.; Pradhan, D. Controlled Synthesis of CuS/TiO_2 Heterostructured Nanocomposites for Enhanced Photocatalytic Hydrogen Generation through Water Splitting. *Inorg. Chem.* **2018**, *57*, 4524–4533. [CrossRef] [PubMed]
57. Jia, S.; Li, X.; Zhang, B.; Yang, J.; Zhang, S.; Li, S.; Zhang, Z. TiO_2/CuS heterostructure nanowire array photoanodes toward water oxidation: The role of CuS. *Appl. Surf. Sci.* **2019**, *463*, 829–837. [CrossRef]
58. Liu, J.; Sun, X.; Fan, Y.; Yu, Y.; Li, Q.; Zhou, J.; Gu, H.; Shi, K.; Jiang, B. P-N Heterojunction Embedded CuS/TiO_2 Bifunctional Photocatalyst for Synchronous Hydrogen Production and Benzylamine Conversion. *Small* **2024**, *20*, e2306344. [CrossRef]
59. Wang, Q.; An, N.; Bai, Y.; Hang, H.; Li, J.; Lu, X.; Liu, Y.; Wang, F.; Li, Z.; Lei, Z. High photocatalytic hydrogen production from methanol aqueous solution using the photocatalysts CuS/TiO_2. *Int. J. Hydrogen Energy* **2013**, *38*, 10739–10745. [CrossRef]
60. Lutterotti, L.; Matthies, S.; Wenk, H.R. MAUD: A friendly Java program for material analysis using diffraction. *IUCr Newsl. Comm. Powder Diffr.* **1999**, *21*, 14–15.
61. Yu, B.; Meng, F.; Zhou, T.; Fan, A.; Khan, M.W.; Wu, H.; Liu, X. Construction of hollow TiO_2/CuS nanoboxes for boosting full-spectrum driven photocatalytic hydrogen evolution and environmental remediation. *Ceram. Int.* **2021**, *47*, 8849–8858. [CrossRef]
62. Zhao, X.; Zhao, K.; Su, J.; Sun, L. TiO_2/CuS core-shell nanorod arrays with aging-induced photoelectric conversion enhancement effect. *Electrochem. Commun.* **2020**, *111*, 106648. [CrossRef]
63. Huang, Z.; Wen, Z.X.; Xiao, X. Photoelectrochemical Properties of $CuS-TiO_2$ Composite Coating Electrode and Its Preparation via Electrophoretic Deposition. *J. Electrochem. Soc.* **2011**, *158*, H1247–H1251. [CrossRef]
64. Tang, Y.; Chai, Y.; Liu, X.; Li, L.; Yang, L.; Liu, P.; Zhou, Y.; Ju, H.; Cheng, Y. A photoelectrochemical aptasensor constructed with core-shell $CuS-TiO_2$ heterostructure for detection of microcystin-LR. *Biosens. Bioelectron.* **2018**, *117*, 224–231. [CrossRef] [PubMed]
65. Wang, Y.; Bai, L.; Wang, Y.; Qina, D.; Shan, D.; Lu, X. Ternary nanocomposites of $Au/CuS/TiO_2$ for ultrasensitive photoelectrochemical non-enzymatic glucose sensor. *Analyst* **2018**, *143*, 1699–1704. [CrossRef] [PubMed]
66. Ngaotrakanwiwat, P.; Heawphet, P.; Rangsunvigit, P. Enhancement of photoelectrochemical cathodic protection of copper in Marine Condition by Cu-Doped TiO_2. *Catalysts* **2020**, *10*, 146. [CrossRef]
67. Liu, W.; Ji, H.; Wang, J.; Zheng, X.; Lai, J.; Ji, J.; Li, T.; Ma, Y.; Li, H.; Zhao, S.; et al. Synthesis and Photo-Response of CuS thin films by an in situ multi-deposition process at room temperature: A Facile & eco-friendly approach. *NANO Brief Rep. Rev.* **2015**, *10*, 1550032.

Disclaimer/Publisher's Note: The statements, opinions and data contained in all publications are solely those of the individual author(s) and contributor(s) and not of MDPI and/or the editor(s). MDPI and/or the editor(s) disclaim responsibility for any injury to people or property resulting from any ideas, methods, instructions or products referred to in the content.

Review

Research Progress on Rashba Effect in Two-Dimensional Organic–Inorganic Hybrid Lead Halide Perovskites

Junhong Guo [1], Jinlei Zhang [2], Yunsong Di [3],* and Zhixing Gan [3,4],*

[1] College of Electronic and Optical Engineering & College of Flexible Electronics (Future Technology), Nanjing University of Posts and Telecommunications, Wenyuan Road 9, Nanjing 210023, China; jhguo@njupt.edu.cn
[2] School of Physical Science and Technology, Suzhou University of Science and Technology, Suzhou 215009, China; zhangjinlei@usts.edu.cn
[3] Center for Future Optoelectronic Functional Materials, School of Computer and Electronic Information, Nanjing Normal University, Nanjing 210023, China
[4] College of Materials Science and Engineering, Qingdao University of Science and Technology, Qingdao 266042, China
* Correspondence: diyunsong@njnu.edu.cn (Y.D.); zxgan@njnu.edu.cn (Z.G.)

Abstract: The Rashba effect appears in the semiconductors with an inversion–asymmetric structure and strong spin-orbit coupling, which splits the spin-degenerated band into two sub-bands with opposite spin states. The Rashba effect can not only be used to regulate carrier relaxations, thereby improving the performance of photoelectric devices, but also used to expand the applications of semiconductors in spintronics. In this mini-review, recent research progress on the Rashba effect of two-dimensional (2D) organic–inorganic hybrid perovskites is summarized. The origin and magnitude of Rashba spin splitting, layer-dependent Rashba band splitting of 2D perovskites, the Rashba effect in 2D perovskite quantum dots, a 2D/3D perovskite composite, and 2D-perovskites-based van der Waals heterostructures are discussed. Moreover, applications of the 2D Rashba effect in circularly polarized light detection are reviewed. Finally, future research to modulate the Rashba strength in 2D perovskites is prospected, which is conceived to promote the optoelectronic and spintronic applications of 2D perovskites.

Keywords: Rashba effect; photoluminescence; 2D perovskites; optoelectronics and spintronics

Citation: Guo, J.; Zhang, J.; Di, Y.; Gan, Z. Research Progress on Rashba Effect in Two-Dimensional Organic–Inorganic Hybrid Lead Halide Perovskites. *Nanomaterials* 2024, 14, 683. https://doi.org/10.3390/nano14080683

Academic Editor: Michael Saliba

Received: 22 March 2024
Revised: 9 April 2024
Accepted: 13 April 2024
Published: 16 April 2024

Copyright: © 2024 by the authors. Licensee MDPI, Basel, Switzerland. This article is an open access article distributed under the terms and conditions of the Creative Commons Attribution (CC BY) license (https://creativecommons.org/licenses/by/4.0/).

1. Introduction

Organic–inorganic hybrid lead halide perovskites (OILHPs) have attracted significant interest in the past years due to their outstanding performance as solar absorbers in photovoltaics [1–5]. The long carrier lifetime of photogenerated carriers is a crucial factor for excellent optoelectronic performance [6]. An extraordinarily long carrier lifetime ($\tau \geq 1$ µs) and a substantial carrier diffusion length ($L_D \geq 5$ µm) have been measured in polycrystalline perovskite thin films with moderate mobility ($\mu \approx 1$–100 cm^2 V^{-1} s^{-1}), which is drastically lower than that of other conventional semiconductors, such as GaAs ($\mu \approx 500$ cm^2 V^{-1} s^{-1}). However, the physical mechanism behind the long carrier lifetime is still elusive [7–11]. The mainstream investigation attributes it to the low trap density [12,13], which may lead to a significant suppression in nonradiative recombination, thus greatly prolonging the carrier's lifetime. However, further research has found that, in perovskites with relatively high defect density, the carrier lifetime does not significantly decrease [14]. Therefore, the correlation between carrier lifetime and defect density in perovskite is not definite. Currently, other models, such as high defect tolerance [15–17], photon recycling [18,19], weak electron–phonon coupling [20–23], the presence of ferroelectric domains [24,25], the formation of polarons, and the screening of band-edge charges [26], have been proposed to rationalize the long carrier lifetime of perovskites. However, after years of laborious exploration, there are still some inherent limitations and inconsistencies in the above-mentioned models.

Among them, the Rashba effect is also considered to be one of the most essential reasons for the long carrier lifetime [27–29]. The Rashba effect was proposed in the 1950s, which reveals spin splitting in noncentrosymmetric semiconductors [30,31]. For ordinary semiconductors, the dispersion of the conduction band minimum (CBM) electrons and valence band maximum (CBM) holes can be described as a spin-degenerate parabolic energy band,

$$E(\mathbf{k}) = \hbar^2 k^2 / 2m^* \tag{1}$$

where k is the electron wavevector, $\hbar$ is the reduced Planck constant, and m^* is the effective mass of electrons (or holes). However, if the semiconductor lacks inversion symmetry, and meanwhile there is strong spin-orbit coupling, an effective magnetic field $\Omega(k)$ appears (Figure 1a), which lifts the degeneracy of the carrier spin states within each band [32]. Thus, when the Rashba effect occurs, the spin-degenerate band splits into two spin-polarized sub-bands deviating from the symmetric center of the Brillouin zone (Figure 1b,c).

$$E_\pm(\mathbf{k}) = \hbar^2 k^2 / 2m^* \pm \alpha_R |\mathbf{k}| \tag{2}$$

α_R is the Rashba splitting constant.

$$\alpha_R = \frac{2E_R}{k_R} \tag{3}$$

Figure 1c shows that the Rashba effect has two important characteristics, namely, energy band splitting and in-plane spin splitting. E_R and k_R are the energy difference and momentum offset between the vertex of the energy curve and the k origin at the high-symmetry point, respectively. The strength of the Rashba effect is usually characterized by the Rashba constant α_R.

Due to the different orbital compositions of the conduction band and valence band, the splitting degrees of the CBM and VBM are not equal. Therefore, the splitting will make the carrier recombination in perovskite exhibit features similar to indirect bandgap, thereby reducing the carrier recombination rate. In addition, because the conduction band and valence band have opposite spin helicity, carrier recombination is spin forbidden, which further reduces the electron-hole recombination rate. Optical selection rules for interband transitions at the band gap are plotted in Figure 1d. The Rashba effect not only provides a possible explanation for the long carrier lifetime in perovskites but also enables effective control and manipulation of the polarized spins in spintronic devices. Apart from the research on conventional optoelectronics areas, such as solar cells, LEDs, and photodetectors [33–36], one of the exciting research directions on lead halide perovskites would be spintronics-related technology.

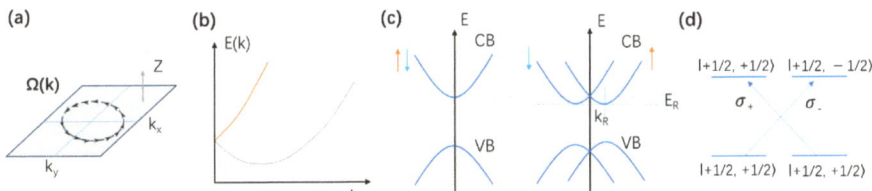

Figure 1. (a) Effective magnetic field $\Omega(\mathbf{k})$ induced by the Rashba effect showing the variation of the direction at a fixed value of $|\mathbf{k}|$. (b) Energies of the spin eigenstates as a function of the in-plane wave vector. (c) The electron dispersion relation shows a doubly degenerate parabolic band at k = 0 subject to Rashba spin splitting, fostering two parabolic bands with opposite spins (arrows). (d) Optical selection rules for interband transitions at the band gap.

The Rashba effect is considered one of the most essential reasons for the long carrier lifetime of OILHPs. Moreover, the photoelectronic properties of OILHPs that can be regulated by the magnitude of the Rashba effect. Thus, the Rashba effect in 2D OILHPs has attracted increasing research interest. In this mini-review, the recent research progress on the Rashba effect in 2D perovskites is summarized. Several important aspects of the Rashba effect in 2D perovskites, including the origin and magnitude, layer dependence in 2D perovskite, the Rashba effect in 2D perovskite quantum dots, 2D/3D composite, and van der Waals heterostructures, are included. In addition, circularly polarized light-detection applications based on the Rashba effect are discussed. Due to the limitation of scope, this review does not include all achievements related to this topic, and only a selection of representative examples is discussed. We hope this mini-review can further stimulate research enthusiasm on this important topic so that more insights into the fundamental understanding can be gained and more optoelectronic and spintronic applications can be developed.

2. Rashba Effect in Three-Dimensional (3D) Perovskites

It is generally believed that the crystal structure of three-dimensional OILHP (such as $MAPbI_3$, MA = methylamine) at room temperature is a tetragonal system (or cubic system or orthogonal system, depending on the material composition) with centrosymmetry. However, it has been found that the perovskite lattice does not have strict centrosymmetry [29]. The lead halide octahedron $[MX_6]^{4-}$ in the perovskite lattice is slightly distorted [37], and the organic cation A^+ also has a certain orientation in a rapidly rotating state [38,39]. These properties may disrupt the centrosymmetry of the perovskite lattice. In addition, there is strong spin-orbit coupling due to the presence of heavy elements, such as lead, tin, and iodine. The Rashba effect in perovskites is expected to be strong. Based on the above reasons, many theoretical studies predict a strong Rashba effect in perovskites [37,40]. For example, the spin-orbit coupling in $MAPbI_3$ causes a displacement of the conduction band energy level of more than 1 eV [41]. In addition, a few experimental studies also strongly support the occurrence of the Rashba effect in the compound [42,43]. A significant effort regarding the experimental observation of Rashba spin-splitting has been demonstrated by Giovanni and co-workers through spin-dependent circularly polarized pump-probe experiments [42]. Neisner et al. directly observed the split in the valence band by angle-resolved photoemission spectroscopy measurements [43].

3. Rashba Effect in Two-Dimensional Perovskites

Two-dimensional OILHPs are commonly known as the Ruddlesden–Popper (RP) phase [44–51] and the Dion–Jacobson (DJ) phase [52]. Taking RP-phase 2D perovskite as an example, its general chemical structure is $(RNH_3)_2A_{n-1}M_nX_{3n-1}$ (n = 1, 2, 3, 4...), where RNH_3 is usually an organic group of aliphatic or aromatic alkylammonium, such as 2-phenylethylammonium (PEA) and n-butylammonium (n-BA), A is a monovalent organic cation, such as $CH_3NH_3^+$ (abbreviated as MA^+) and $HC(NH_2)_2^+$ (abbreviated as FA^+), and M is a divalent metal cation, mainly referring to lead Pb, X is a halide anion. The large organic cations (RNH_3^+) separate the layers of the inorganic Pb-I network. And n represents the number of inorganic $[MX_6]^{4-}$ octahedral structures in each period. The 2D OILHPs have attracted increasing research interest due to their special multi-quantum-well structures and excellent structural stability under ambient conditions [53,54].

Bychkov and Rashba proposed that the Rashba effect also appears in two-dimensional (2D) electron gas systems [55]. Thus, the Rashba effect has been extensively investigated in various 2D material systems, including III−V semiconductor heterostructures and topological insulators Bi_2Se_3 over the past few decades [56–58]. Nevertheless, Rashba splitting energy in these 2D structures is typically smaller than 10 meV, limiting the performance of spintronic devices based on these 2D materials [56–58]. The Rashba effect in 2D perovskites has also attracted extensive research attention. Density functional theory (DFT) calculation is an important tool for defining and demonstrating the existence of Rashba splitting, as well as quantifying the structural symmetry, rotation, and distortion. For example, Zhai et al. showed the existence of Rashba splitting in the plane perpendicular to the 2D layer of $(C_6H_5C_2H_4NH_3)_2PbI_4$ based on the DFT calculations using local density approximation (LDA) in the form of ultrasoft pseudopotentials [59]. In more detail, the first-principles DFT calculations show that the breaking of inversion symmetry is caused by the displacement of the Pb atom from the octahedra center, which leads to the Rashba splitting. At temperatures below 110 K, the absorption spectrum in the photon energy range of 2.45 to 2.65 eV shows two step-like absorption edges, which are assigned as the 1s and 2s exciton energy at 2.38 and 2.53 eV, respectively (Figure 2a). Considering the band edge of $(C_6H_5C_2H_4NH_3)_2PbI_4$ at 2.57 eV (Figure 2b), 1s and 2s exciton binding energies are about 190 ± 4 meV and 45 ± 8 meV, respectively. The energy differences between δE_{ac} and δE_{bc} scales with $V^{2/3}$ indicate a Frank–Keldysh-type oscillatory feature at the continuum band edge (Figure 2c). From the electroabsorption spectrum and photoinduced absorption spectra of excitons and free carriers, they obtained a giant Rashba splitting in 2D $(C_6H_5C_2H_4NH_3)_2PbI_4$ thin film, with energy splitting of (40 ± 5) meV and a Rashba constant of (1.6 ± 0.1) eV·Å (Figure 2d) [59].

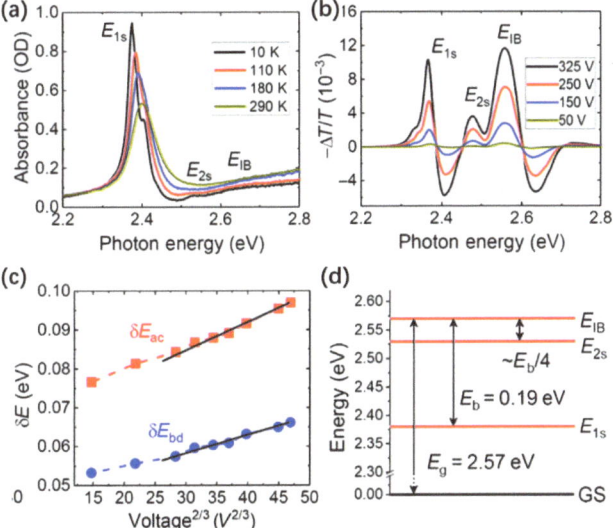

Figure 2. (a) Absorption spectra of $(C_6H_5C_2H_4NH_3)_2PbI_4$ film at various temperatures, which contains 1s and 2s exciton (labeled as E_{1s} and E_{2s}, respectively) and an interband (IB) transition. (b) Electroabsorption spectra of $(C_6H_5C_2H_4NH_3)_2PbI_4$ thin film measured at 45 K at various applied to electric fields. (c) Energy differences δE_{ac} and δE_{bc} plotted versus $V^{2/3}$. (d) Energy levels of the excitons and interband transition (IB) with respect to the ground state (GS) [59]. Reproduced with permission under Creative Common CC-BY 4.0 license.

In addition, Todd et al. investigated carrier dynamics in 2D $(BA)_2MAPb_2I_7$ thin film by time-resolved circular dichroism techniques [60]. They revealed the presence of a Rashba spin splitting via the dominance of processional spin relaxation induced by the Rashba effective magnetic field. The Rashba spin-splitting magnitude was extracted from simulations of the measured spin dynamics incorporating longitudinal optical-phonon and electron–electron scattering, yielding a value of 10 meV at an electron energy of 50 meV above the band gap, which is twenty times larger than that in GaAs quantum wells. Moreover, a Rashba splitting of 85 meV with a Rashba coefficient α_R of 2.6 eV Å was observed in an emergent 2D DJ phase $(AMP)PbI_4$ (AMP = 4-(aminomethyl)piperidinium) [61]. Jana et al. introduced a structural chirality transfer across the organic–inorganic interface in 2D perovskites using appropriate chiral organic cations [62]. The chiral spacer cations and their asymmetric hydrogen-bonding interactions with lead bromide-based layers cause symmetry-breaking helical distortions in the inorganic layers. The first-principles calculation predicts a substantial bulk of the Rashba–Dresselhaus spin-splitting in the inorganic-derived conduction band with opposite spin textures between R- and S-hybrids due to the broken inversion symmetry and strong spin-orbit coupling. The chirality transfer from one structural unit to another represents a promising approach to breaking symmetry that modulates the Rashba effect for spintronics and related applications. These findings indicated that 2D hybrid perovskites have great potential for applications in spintronics.

3.1. Origin and Magnitude of Rashba Spin Splitting in 2D RP Perovskites

Rashba spin splitting has been observed in multiple 2D OILHPs, yet with a significant variance in the magnitude of spin splitting [58–61]. However, the origin of the giant Rashba splitting remains elusive. The crucial role of the orientation of the organic cation in the 2D RP perovskite was explored by Kagdada et al. Their DFT calculation results revealed that the MA cation rotation imposes structural distortion in the inorganic PbI_6 layer, which then varies the structure and value of the electronic bandgap, charge density, and optical absorption. The strong spin–orbit coupling leads to a wide range of Rashba splitting parameters from 0.04 to 0.278 eV Å. The simulated optical absorption spectra showed that absorption edges for the different orientations of the MA molecule are not the same [63].

In addition, Zhou et al. obtained (AMP)PbI_4 DJ phase crystals by an economical aqueous method. They clarified the origin of the giant Rashba effect by temperature- and polarization-dependent photoluminescence (PL) results [64]. The strong temperature-dependent PL helicity indicates the thermally assisted structural distortion as the main origin of the Rashba effect, suggesting that valley polarization still preserves at high temperatures. The Rashba effect was further confirmed by the circular photogalvanic effect near the indirect bandgap (Figure 3).

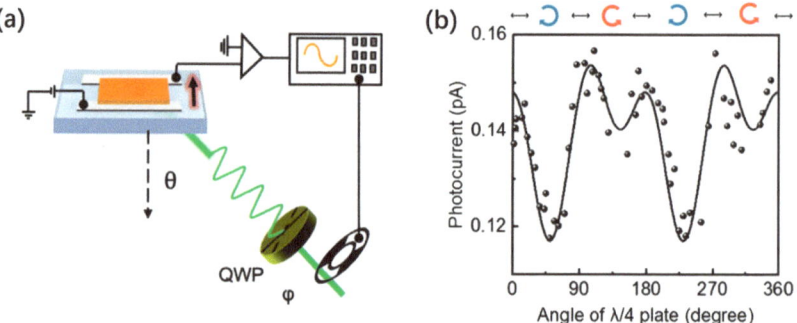

Figure 3. (a) Schematic illustration of the experimental setup for measurement of photogalvanic current. The φ indicates the angle between the fast axis of the quarter-wave plate (QWP) and the incident light polarization. The θ indicates the incident angle of excitation light. (b) Room temperature photogalvanic current of (AMP)PbI_4 versus QWP rotation angle φ, measured at $\theta = 60°$ and excited via a 556 nm continuous laser [64]. Reproduced with permission. Copyright 2021, American Chemical Society.

In addition, organic–inorganic hybrid halide perovskites are susceptible to dynamic instabilities known as octahedral tilt, which involves a rigid rotation of the inorganic octahedral cages and can occur along any of the three Cartesian directions in the crystal with either in-phase or out-of-phase ordering [65]. While the phase transitions related to octahedral tilt have been thoroughly examined in 3D hybrid halide perovskites, their influence on hybrid 2D perovskites remains not fully comprehended. To gain insight into this puzzle, Shao et al. utilized scanning tunneling microscopy to directly visualize the surface octahedral tilt in freshly exfoliated 2D RP perovskites across the homologous series [66]. The steric hindrance imposed by long organic cations is unlocked by exfoliation. The experimentally determined octahedral tilts from 2D RP-phase perovskites of $n = 1$ to $n = 4$ align closely with the out-of-plane surface octahedral tilts predicted by DFT calculations. The out-of-plane octahedral tilt of the exfoliated surface is correlated to the redshifted emission peak alongside the primary exciton in the PL spectra. Therefore, the Rashba spin splitting is attributed to the octahedral tilt [66].

3.2. Layer-Dependent Rashba Band Splitting in 2D Perovskites

It is very significant to reveal the impacts of surface termination and the number of inorganic layers on the amplitude of Rashba band splitting so as to enhance the understanding of the origin and extent of Rashba spin splitting in 2D RP-phase perovskites. Thus, research efforts were devoted to the layer-dependent Rashba band splitting in 2D perovskites. Singh et al. investigated Rashba spin splitting in 2D RP $(BA)_2(MA)_{n-1}Pb_nI_{3n+1}$ with both centrosymmetric ($n = 1$) and noncentrosymmetric ($n = 2$ and 3) structures, using first-principle calculations, polarization, and temperature-dependent PL spectroscopy [67]. They revealed the n-dependent Rashba spin splitting in 2D RP perovskites. When $n = 1$, a single metal halide octahedral layer is sandwiched between long BA^+ organic cations, Rashba spin splitting is the largest. As n increases, the Rashba spin splitting decreases. The large Rashba effect observed in the 2D RP perovskite of an $n = 1$ structure is attributed to the local distortion of the PbI_6 octahedron at the surface [67].

By using a combination of DFT calculations and time-resolved PL spectroscopy, Yin et al. compared the Rashba band splitting of the prototype 3D MAPbI$_3$ and the 2D RP perovskites [68]. They demonstrated that significant structural distortions associated with different surface terminations are responsible for the observed Rashba effect in 2D OILHPs. Interestingly, their calculation results indicated that the intrinsic Rashba splitting occurs in the perovskite crystals with an even number of inorganic layers ($n = 2$), in consistency with their longer PL lifetimes and ground-state bleaching recovery lifetimes. Whereas, when the number of inorganic layers is odd ($n = 1$ and $n = 3$), the Rashba effect of 2D RP perovskites absences (Figure 4). These findings elucidate the significant impact of the number of inorganic layers on the electronic properties of 2D perovskites, suggesting the controlling of the n value in 2D RP perovskites to design Rahsba effects for spintronic applications.

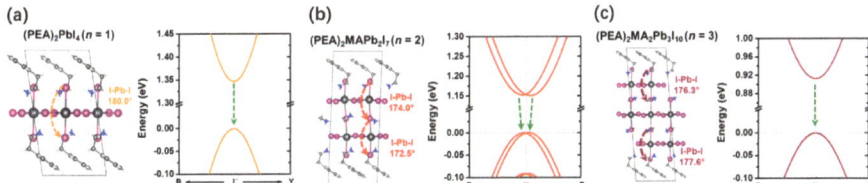

Figure 4. (a) Optimized crystal structures and electronic band structures of (a) (PEA)$_2$PbI$_4$ ($n = 1$), (b) (PEA)$_2$MAPb$_2$I$_7$ ($n = 2$), and (c) (PEA)$_2$MA$_2$Pb$_3$I$_{10}$ ($n = 3$) [68]. Reproduced with permission. Copyright 2018, American Chemical Society.

In addition, Liu investigated the thickness-dependent structural distortion along with the Rashba splitting energy by using the DFT calculation [69]. Three types of OILHPs were compared to explore the effect of halogens and organic ligands. As the thickness increases, the structural distortion degree decreases. The Rashba splitting magnitude follows the same tendency. The 2D MAPbI$_3$ is less sensitive to thickness change compared to the 2D MAPbBr$_3$ or the 2D MAPbCl$_3$. Furthermore, ligands and their orientations have dramatically different impacts on the Rashba splitting. The PEA ligands enhance the Rashba splitting magnitude, while the BA ligands have the converse effect. The partial charge-density analysis shows that the band edges are contributed to by a charge density at a specific layer in the structure. Thus, they concluded that the Rashba effect is layer dependent in 2D HOIPs [69].

3.3. Rashba Effect in 2D Perovskite Quantum Dots

Because of the quantum confinement effect, the quantum dot usually shows fast radiative recombination, large exciton binding energies [70], and giant oscillator transition strengths [71]. Most theoretical descriptions of the Rashba effect on exciton fine structures were conducted in the weak-confinement regime, in which the exciton Bohr radius, r_B, is much smaller than the typical size of the nanocrystals. The Rashba effect was treated perturbatively, which is a valid approach, assuming $\alpha k \ll \frac{\hbar^2 k^2}{2m^*}$, where α_e and α_h are the Rashba coefficients in the conduction and valence bands, respectively, k is the typical quasi-momentum of exciton center-of-mass (COM) motion, and m^* is the effective mass of the COM motion. The momentum is $k \sim 1/R$ for an exciton confined in an NC with size R, so the perturbative approach is valid when $\alpha \ll \hbar^2/2m^*R$. This condition is clearly not satisfied in a large NC ($R \gg \hbar^2/2m^*\alpha$) or in NCs with enormously large Rashba coefficients. Thus, the Rashba effect in 2D perovskite quantum dots is elusive. To explore this question, Swift et al. constructed an effective mass model of excitons in 2D perovskite quantum dots, which covers the full range of NC sizes and Rashba strengths [72]. The fine structure and oscillator transition strengths of Rashba excitons confined in a 2D cylindrical quantum dot are quite unusual. One notable aspect of the energy-level structure is the proliferation of dark exciton states. These dark states in large quantum dots are also likely to be thermally populated even at quite low temperatures, reducing the radiative decay rate and, consequently, the PL quantum yield of these structures.

3.4. Rashba Effect in 2D/3D Composite Perovskite Films

Compared with common 2D perovskite, the 2D/3D composite perovskite may have a variety of gains, such as significant interface asymmetry and an effective energy-transfer process. On the one hand, the interface asymmetry can enhance the band splitting. On the other hand, energy transfer can be used to improve the photoresponse. These two effects make 2D/3D composite perovskite promising for opto-spintronic applications. The recent development of chiral 2D/three-dimensional (3D) composite perovskites offers a new opportunity to engineer the Rashba effect. Li et al. synthesized one pair of chiral 2D/3D composite perovskite [73]. The optical properties were studied by polarization-dependent femtosecond transient absorption (fs-TA) spectroscopy, which revealed that the chiral properties of organic cations were successfully transferred to the achiral part. The Rashba effect is significantly enhanced in the 2D/3D composite structures. The spintronic relaxation along with the Rashba effect in the 2D/3D composite structures will inspire the further development of the next generation of opto-spintronic devices.

3.5. Rashba Effect of Van Der Waals Heterostructures Based on 2D Perovskites

The van der Waals heterostructures based on different 2D materials enable innovative device engineering. A variety of van der Waals heterostructures have been developed based on 2D perovskites for optoelectronic applications. Thus, it is very significant to investigate the Rashba effect in van der Waals heterostructures. Singh et al. integrate an RP-phase 2D perovskites monolayer with another important family of 2D excitonic semiconductors, i.e., transition-metal dichalcogenides (TMDs) [67]. A combined effect of Rashba spin splitting in 2D RP perovskites and the strong spin–valley physics of monolayer TMDs can give rise to effective spin–valley polarization in the heterostructures using circularly polarized light (CPL) excitation. Thus, the 2D RP perovskite/TMD heterostructure provides an attractive material combination for investigating valleytronic phenomena, as it reduces fabrication complexity and sample-to-sample variance. Different 2D RP perovskites ($n = 1$ and 2) and monolayer WSe$_2$s were coupled to form 2D vdW heterostructures. Robust interlayer excitons (IXs) in staggered type-II band-aligned heterostructures were observed (Figure 5). These IXs are strongly valley-polarized with exciton lifetimes longer than the intralayer excitons in the constituent monolayer TMDs, suggesting the spin–valley-dependent optical selection rules to the IXs. This research broadens the scope for exploring spin–valley physics in heterogeneous stacks of 2D semiconductors. They also investigated a 2DRP-(n = 1)/MoS$_2$ heterostructure with a broken type-III band alignment. In contrast, there is no interlayer charge transfer, thus the 2DRP/MoS$_2$ heterostructure does not show any IX emission.

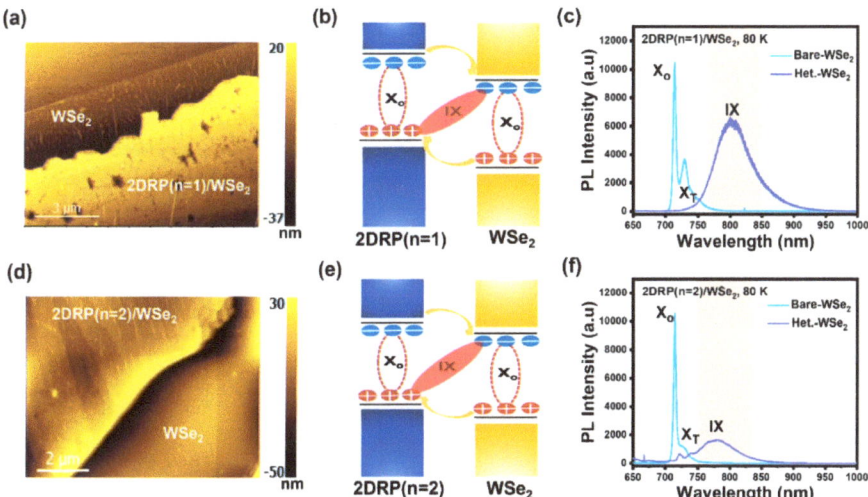

Figure 5. Topography (**a,d**), band structures (**b,e**), and PL spectra (**c,f**) of 2D RP (n = 1)/WSe$_2$ (**a–c**) and 2D RP (n = 2)/WSe$_2$ heterostructures (**d–f**) under 633 nm excitation. The type-II band alignment between monolayer WSe$_2$ and 2D RP perovskites leads to the formation of interlayer excitons [67]. Reproduced with permission. Copyright 2023, American Chemical Society.

3.6. Applications of 2D Rashba Effect in Circularly Polarized Light Detection

CPL is a special light beam, which consists of two spiral modes called chirality or handedness. Based on the rotation of the field vector, the CPL can either rotate counterclockwise (left handed, σ^+) or clockwise (right handed, σ^-) when observed from the direction opposite to the wave's propagation. Direct detection of CPL is a challenging task due to limited materials and ambiguous structure–property relationships that lead to low distinguishability of the light helicities. On the one hand, the circular photogalvanic effect is considered the most important experiment that confirms the presence of the Rashba effect in semiconductors. The circular photogalvanic effect has been demonstrated in a variety of materials with the Rashba effect, such as GaAs/AlGaAs multi-quantum wells, the polar semiconductor BiTeI, 2D transition-metal dichalcogenides, and topological insulators [74–77]. On the other hand, the Rashba effect in 2D perovskites provides new opportunities for dealing with the challenge of CPL detection.

Chiral 2D perovskites have been recently explored as the responsive component for the direct detection of CPL [78–84]. For example, Wang et al. inserted chiral organic ligands into the organic layers of 2D perovskites to obtain chiral (R-MBA)$_2$PbI$_4$ and (S-MBA)$_2$PbI$_4$. The in-plane photocurrent response generated by the CPL excitation of planar photoconductive devices shows a typical response of the chirality-induced circular photogalvanic effect that originates from the Rashba splitting in the electronic bands of these compounds, demonstrating the potential applications of chiral 2D perovskites in optoelectronic devices that are sensitive to the light helicity [85]. Similarly, Fan et al. report direct CPL detection by using a pair of 2D chiral perovskite ferroelectrics, (R/S-3AMP)PbBr$_4$ (3AMP = 3-(aminomethyl)-piperidine divalent cation) [86]. These 2D perovskites undergo a phase transition at 420 K that is a combination of order–disorder and displacive ferroelectric transition. DFT calculations and circularly polarized light-excited PL measurements have confirmed the presence of the Rashba effect in these 2D chiral perovskites (Figure 6a–d). This effect results in spin selectivity, which can modulate the behavior of photogenerated charge carriers during transitions, recombination, and transfers. Single-crystal-based devices have been shown to directly detect CPL at 430 nm, with an on–off ratio of current higher than 1.7×10^3 and anisotropy factors of responsivity larger than 0.20 (Figure 6e). The enhanced CPL detection is attributed to the Rashba effect, which has a large Rashba coefficient of 0.93 eV·Å.

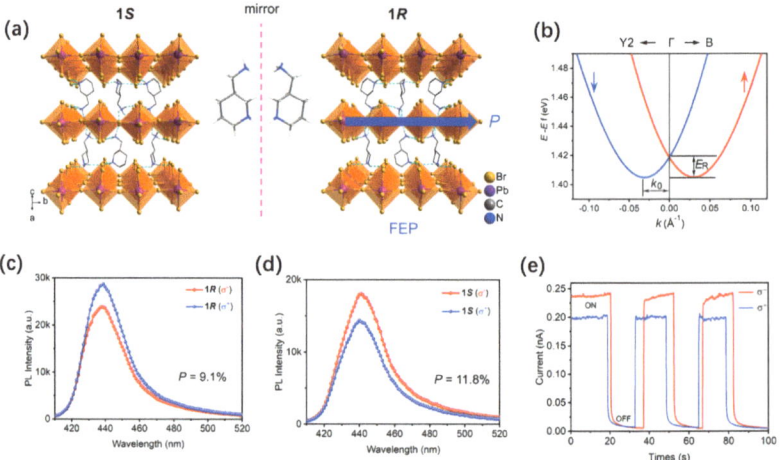

Figure 6. (a) Crystal structures of (S-3AMP)PbBr$_4$ (1S) and (R-3AMP)PbBr$_4$ (1R) in the ordered ferroelectric phase (FEP). (b) Rashba splitting band structure of 1R. (c,d) CPLEPL spectra of 1R (c) and 1S (d) upon L-CPL (σ^+) and R-CPL (σ^-) excitation at 395 nm. (e) Photocurrent differences upon L- and R-CPL irradiation at 430 nm [86]. Reproduced with permission. Copyright 2022, Wiley-VCH.

4. Conclusions and Outlook

In summary, this mini-review focuses on the Rashba effect in 2D perovskites. Recent research progress on the origin and extent of Rashba spin splitting, layer-dependent Rashba band splitting of 2D perovskites, the Rashba effect on 2D perovskite quantum dots, the Rashba effect in 2D/3D composite perovskite, and the Rashba effect in van der Waals heterostructures based on 2D perovskites are reviewed. In addition, applications of the 2D Rashba effect in circularly polarized light detection are included in this review.

Despite considerable reports on Rashba effects in 2D perovskites, the origin of Rashba spin splitting in 2D perovskites is still under debate. Future research efforts to investigate the impacts of the surface termination, the number of inorganic layers, the structure of organic spacers, the planar sizes, and the distortion of inorganic octahedrons on the magnitude of Rashba band splitting will not only gain more insight into the origin of Rashba effect in 2D perovskites but also inspire approaches to modulate the Rashba spin splitting. In addition, the relationship between charge-carrier dynamics and the Rashba effect in 2D perovskites is still to be established, so that the photoelectronic properties and photophysics of 2D perovskites can be effectively controlled by modulating the Rashba magnitude. Apart from the research on conventional optoelectronics areas, such as solar cells, LEDs, and photodetectors, one of the exciting research interests on 2D perovskites will be focused on spintronics-related technology. However, the current related research is still insufficient. In other words, there is plenty of room to design new spintronic devices based on 2D perovskites.

Author Contributions: Conceptualization, Z.G. and J.G.; writing—original draft preparation, J.G. and J.Z.; writing—review and editing, Z.G. and Y.D. All authors have read and agreed to the published version of the manuscript.

Funding: This work was supported by the Natural Science Foundation of Shandong Province (ZR2021YQ32), the China Postdoctoral Science Foundation (2023M740472), and the Taishan Scholars Program of Shandong Province (tsqn201909117), the National Natural Science Foundation of China (62004136).

Conflicts of Interest: The authors declare no conflict of interest.

References

1. Wang, F.; Chang, Q.; Yun, Y.; Liu, S.; Liu, Y.; Wang, J.; Fang, Y.; Cheng, Z.; Feng, S.; Yang, L.; et al. Hole-Transporting Low-Dimensional Perovskite for Enhancing Photovoltaic Performance. *Research* **2021**, *2021*, 9797053. [CrossRef] [PubMed]
2. Gottesman, R.; Zaban, A. Perovskites for Photovoltaics in the Spotlight: Photoinduced Physical Changes and Their Implications. *Acc. Chem. Res.* **2016**, *49*, 320–329. [CrossRef] [PubMed]
3. Huang, J.; Yuan, Y.; Shao, Y.; Yan, Y. Understanding the physical properties of hybrid perovskites for photovoltaic applications. *Nat. Rev. Mater.* **2017**, *2*, 17042. [CrossRef]
4. Aydin, E.; Allen, T.G.; De Bastiani, M.; Razzaq, A.; Xu, L.; Ugur, E.; Liu, J.; De Wolf, S. Pathways toward commercial perovskite/silicon tandem photovoltaics. *Science* **2024**, *383*, eadh3849. [CrossRef] [PubMed]
5. Balaguera, E.H.; Bisquert, J. Accelerating the Assessment of Hysteresis in Perovskite Solar Cells. *ACS Energy Lett.* **2024**, *9*, 478–486. [CrossRef] [PubMed]
6. Maiti, A.; Pal, A.J. Carrier recombination in CH3NH3PbI3: Why is it a slow process? *Rep. Prog. Phys.* **2022**, *85*, 024501. [CrossRef] [PubMed]
7. Ambrosio, F.; Wiktor, J.; De Angelis, F.; Pasquarello, A. Origin of low electron–hole recombination rate in metal halide perovskites. *Energy Environ. Sci.* **2018**, *11*, 101–105. [CrossRef]
8. Chen, T.; Chen, W.-L.; Foley, B.J.; Lee, J.; Ruff, J.P.C.; Ko, J.Y.P.; Brown, C.M.; Harriger, L.W.; Zhang, D.; Park, C.; et al. Origin of long lifetime of band-edge charge carriers in organic–inorganic lead iodide perovskites. *Proc. Natl. Acad. Sci. USA* **2017**, *114*, 7519–7524. [CrossRef] [PubMed]
9. Frost, J.M.; Butler, K.T.; Brivio, F.; Hendon, C.H.; van Schilfgaarde, M.; Walsh, A. Atomistic Origins of High-Performance in Hybrid Halide Perovskite Solar Cells. *Nano Lett.* **2014**, *14*, 2584–2590. [CrossRef]
10. Zhang, C.; Sun, D.; Vardeny, Z.V. Multifunctional Optoelectronic–Spintronic Device Based on Hybrid Organometal Trihalide Perovskites. *Adv. Electron. Mater.* **2017**, *3*, 1600426. [CrossRef]
11. Zhang, Z.; Long, R.; Tokina, M.V.; Prezhdo, O.V. Interplay between Localized and Free Charge Carriers Can Explain Hot Fluorescence in the CH3NH3PbBr3 Perovskite: Time-Domain Ab Initio Analysis. *J. Am. Chem. Soc.* **2017**, *139*, 17327. [CrossRef] [PubMed]
12. Yamada, Y.; Nakamura, T.; Endo, M.; Wakamiya, A.; Kanemitsu, Y. Photocarrier Recombination Dynamics in Perovskite $CH_3NH_3PbI_3$ for Solar Cell Applications. *J. Am. Chem. Soc.* **2014**, *136*, 11610. [CrossRef] [PubMed]
13. Stranks, S.D.; Burlakov, V.M.; Leijtens, T.; Ball, J.M.; Goriely, A.; Snaith, H.J. Recombination Kinetics in Organic-Inorganic Perovskites: Excitons, Free Charge, and Subgap States. *Phys. Rev. Appl.* **2014**, *2*, 034007. [CrossRef]
14. Azarhoosh, P.; McKechnie, S.; Frost, J.M.; Walsh, A.; van Schilfgaarde, M. Research Update: Relativistic origin of slow electron-hole recombination in hybrid halide perovskite solar cells. *APL Mater.* **2016**, *4*, 091501. [CrossRef]
15. Steirer, K.X.; Schulz, P.; Teeter, G.; Stevanovic, V.; Yang, M.; Zhu, K.; Berry, J.J. Defect Tolerance in Methylammonium Lead Triiodide Perovskite. *ACS Energy Lett.* **2016**, *1*, 360–366. [CrossRef]
16. Johnston, M.B.; Herz, L.M. Hybrid Perovskites for Photovoltaics: Charge-Carrier Recombination, Diffusion, and Radiative Efficiencies. *Acc. Chem. Res.* **2015**, *49*, 146–154. [CrossRef]
17. Brandt, R.E.; Poindexter, J.R.; Gorai, P.; Kurchin, R.C.; Hoye, R.L.Z.; Nienhaus, L.; Wilson, M.W.B.; Polizzotti, J.A.; Sereika, R.; Zaltauskas, R.; et al. Searching for "Defect-Tolerant" Photovoltaic Materials: Combined Theoretical and Experimental Screening. *Chem. Mater.* **2017**, *29*, 4667–4674. [CrossRef]
18. Pazos-Outón, L.M.; Szumilo, M.; Lamboll, R.; Richter, J.M.; Crespo-Quesada, M.; Abdi-Jalebi, M.; Beeson, H.J.; Vručinić, M.; Alsari, M.; Snaith, H.J.; et al. Photon recycling in lead iodide perovskite solar cells. *Science* **2016**, *351*, 1430–1433. [CrossRef] [PubMed]
19. Gan, Z.; Wen, X.; Chen, W.; Zhou, C.; Yang, S.; Cao, G.; Ghiggino, K.P.; Zhang, H.; Jia, B. The dominant energy transport pathway in halide perovskites: Photon recycling or carrier diffusion? *Adv. Energy Mater.* **2019**, *9*, 1900185. [CrossRef]
20. Wright, A.D.; Verdi, C.; Milot, R.L.; Eperon, G.E.; Pérez-Osorio, M.A.; Snaith, H.J.; Giustino, F.; Johnston, M.B.; Herz, L.M. Electron–phonon coupling in hybrid lead halide perovskites. *Nat. Commun.* **2016**, *7*, 11755. [CrossRef]
21. Zhao, T.Q.; Shi, W.; Xi, J.Y.; Wang, D.; Shuai, Z.G. Intrinsic and Extrinsic Charge Transport in $CH_3NH_3PbI_3$ Perovskites Predicted from First-Principles. *Sci. Rep.* **2016**, *6*, 19968. [CrossRef] [PubMed]
22. Kirchartz, T.; Markvart, T.; Rau, U.; Egger, D.A. Impact of Small Phonon Energies on the Charge-Carrier Lifetimes in Metal-Halide Perovskites. *J. Phys. Chem. Lett.* **2018**, *9*, 939–946. [CrossRef] [PubMed]
23. Motta, C.; Sanvito, S. Electron–Phonon Coupling and Polaron Mobility in Hybrid Perovskites from First Principles. *J. Phys. Chem. C* **2018**, *122*, 1361–1366. [CrossRef]

24. Frost, J.M.; Butler, K.T.; Walsh, A. Molecular ferroelectric contributions to anomalous hysteresis in hybrid perovskite solar cells. *APL Mater.* **2014**, *2*, 081506. [CrossRef]
25. Liu, S.; Zheng, F.; Koocher, N.Z.; Takenaka, H.; Wang, F.G.; Rappe, A.M. Ferroelectric Domain Wall Induced Band Gap Reduction and Charge Separation in Organometal Halide Perovskites. *J. Phys. Chem. Lett.* **2015**, *6*, 693–699. [CrossRef] [PubMed]
26. Zhu, X.-Y.; Podzorov, V. Charge Carriers in Hybrid Organic–Inorganic Lead Halide Perovskites Might Be Protected as Large Polarons. *J. Phys. Chem. Lett.* **2015**, *6*, 4758–4761. [CrossRef] [PubMed]
27. Zheng, F.; Tan, L.Z.; Liu, S.; Rappe, A.M. Rashba spin–orbit coupling enhanced carrier lifetime in $CH_3NH_3PbI_3$. *Nano Lett.* **2015**, *15*, 7794. [CrossRef] [PubMed]
28. Myung, C.W.; Javaid, S.; Kim, K.S.; Lee, G. Rashba–Dresselhaus effect in inorganic/organic lead iodide erovskite interfaces. *ACS Energy Lett.* **2018**, *3*, 1294. [CrossRef]
29. Zhang, X.; Shen, J.-X.; Van de Walle, C.G. Three-Dimensional Spin Texture in Hybrid Perovskites and Its Impact on Optical Transitions. *J. Phys. Chem. Lett.* **2018**, *9*, 2903–2908. [CrossRef]
30. Dresselhaus, G.; Kip, A.F.; Kittel, C. Spin-Orbit Interaction and the Effective Masses of Holes in Germanium. *Phys. Rev. B* **1954**, *95*, 568–569. [CrossRef]
31. Rashba, E.I. Properties of semiconductors with an extremum loop. 1. Cyclotron and combinational resonance in a magnetic field perpendicular to the plane of the loop. *Phys. Solid State* **1960**, *2*, 1224–1238.
32. Awschalom, D.D.; Loss, D.; Samarth, N. (Eds.) *Semiconductor Spintronics and Quantum Computation*; Springer: Berlin/Heidelberg, Germany; New York, NY, USA, 2002.
33. Protesescu, L.; Yakunin, S.; Bodnarchuk, M.I.; Krieg, F.; Caputo, R.; Hendon, C.H.; Yang, R.X.; Walsh, A.; Kovalenko, M.V. Nanocrystals of Cesium Lead Halide Perovskites ($CsPbX_3$, X = Cl, Br, and I): Novel Optoelectronic Materials Showing Bright Emission with Wide Color Gamut. *Nano Lett.* **2015**, *15*, 3692–3696. [CrossRef] [PubMed]
34. Triana, M.A.; Hsiang, E.-L.; Zhang, C.; Dong, Y.; Wu, S.-T. Luminescent Nanomaterials for Energy-Efficient Display and Healthcare. *ACS Energy Lett.* **2022**, *7*, 1001–1020. [CrossRef]
35. Zhang, C.; He, Z.; Chen, H.; Zhou, L.; Tan, G.; Wu, S.-T.; Dong, Y. Light diffusing, down-converting perovskite-on-polymer microspheres. *J. Mater. Chem. C* **2019**, *7*, 6527–6533. [CrossRef]
36. Feng, X.; Sheng, Y.; Ma, K.; Xing, F.; Liu, C.; Yang, X.; Qian, H.; Zhang, S.; Di, Y.; Liu, Y.; et al. Multi-Level Anti-Counterfeiting and Optical Information Storage Based on Luminescence of Mn-Doped Perovskite Quantum Dots. *Adv. Opt. Mater.* **2022**, *10*, 2200706. [CrossRef]
37. Kim, M.; Im, J.; Freeman, A.J.; Ihm, J.; Jin, H. Switchable S = 1/2 and J = 1/2 Rashba bands in ferroelectric halide perovskites. *Proc. Natl. Acad. Sci. USA* **2014**, *111*, 6900. [CrossRef] [PubMed]
38. Etienne, T.; Mosconi, E.; De Angelis, F. Dynamical Rashba Band Splitting in Hybrid Perovskites Modeled by Local Electric Fields. *J. Phys. Chem. C* **2018**, *122*, 124–132. [CrossRef]
39. Hu, S.; Gao, H.; Qi, Y.; Tao, Y.; Li, Y.; Reimers, J.R.; Bokdam, M.; Franchini, C.; Di Sante, D.; Stroppa, A.; et al. Dipole Order in Halide Perovskites: Polarization and Rashba Band Splittings. *J. Phys. Chem. C* **2017**, *121*, 23045–23054. [CrossRef]
40. Stroppa, A.; Di Sante, D.; Barone, P.; Bokdam, M.; Kresse, G.; Franchini, C.; Whangbo, M.-H.; Picozzi, S. Tunable ferroelectric polarization and its interplay with spin–orbit coupling in tin iodide perovskites. *Nat. Commun.* **2014**, *5*, 5900. [CrossRef]
41. Manser, J.S.; Christians, J.A.; Kamat, P.V. Intriguing Optoelectronic Properties of Metal Halide Perovskites. *Chem. Rev.* **2016**, *116*, 12956–13008. [CrossRef]
42. Giovanni, D.; Ma, H.; Chua, J.; Gratzel, M.; Ramesh, R.; Mhaisalkar, S.; Mathews, N.; Sum, T.C. Highly spin-polarized carrier dynamics and ultra large photoinduced magnetization in $CH_3NH_3PbI_3$ perovskite thin films. *Nano Lett.* **2015**, *15*, 1553. [CrossRef] [PubMed]
43. Niesner, D.; Wilhelm, M.; Levchuk, I.; Osvet, A.; Shrestha, S.; Batentschuk, M.; Brabec, C.; Fauster, T. Giant Rashba splitting in $CH_3NH_3PbBr_3$ organic–inorganic perovskite. *Phys. Rev. Lett.* **2016**, *117*, 126401. [CrossRef] [PubMed]
44. Wang, Y.; Zha, Y.; Bao, C.; Hu, F.; Di, Y.; Liu, C.; Xing, F.; Xu, X.; Wen, X.; Gan, Z.; et al. Monolithic 2D Perovskites Enabled Artificial Photonic Synapses for Neuromorphic Vision Sensors. *Adv. Mater.* **2024**, *36*, e2311524. [CrossRef] [PubMed]
45. Zha, Y.; Wang, Y.; Sheng, Y.; Zhang, X.; Shen, X.; Xing, F.; Liu, C.; Di, Y.; Cheng, Y.; Gan, Z. Stable and broadband photodetectors based on 3D/2D perovskite heterostructures with surface passivation. *Appl. Phys. Lett.* **2022**, *121*, 191904. [CrossRef]
46. Chen, Y.; Sun, Y.; Peng, J.; Tang, J.; Zheng, K.; Liang, Z. 2D Ruddlesden–Popper Perovskites for Optoelectronics. *Adv. Mater.* **2018**, *30*, 1703487. [CrossRef] [PubMed]
47. Dou, L.; Wong, A.B.; Yu, Y.; Lai, M.; Kornienko, N.; Eaton, S.W.; Fu, A.; Bischak, C.G.; Ma, J.; Ding, T.; et al. Atomically thin two-dimensional organic-inorganic hybrid perovskites. *Science* **2015**, *349*, 1518–1521. [CrossRef] [PubMed]
48. Blancon, J.-C.; Tsai, H.; Nie, W.; Stoumpos, C.C.; Pedesseau, L.; Katan, C.; Kepenekian, M.; Soe, C.M.M.; Appavoo, K.; Sfeir, M.Y.; et al. Extremely efficient internal exciton dissociation through edge states in layered 2D perovskites. *Science* **2017**, *355*, 1288–1292. [CrossRef] [PubMed]
49. Tsai, H.; Nie, W.; Blancon, J.-C.; Stoumpos, C.C.; Asadpour, R.; Harutyunyan, B.; Neukirch, A.J.; Verduzco, R.; Crochet, J.J.; Tretiak, S.; et al. High-efficiency two-dimensional Ruddlesden–Popper perovskite solar cells. *Nature* **2016**, *536*, 312–316. [CrossRef] [PubMed]
50. Kumagai, M.; Takagahara, T. Excitonic and nonlinear-optical properties of dielectric quantum-well structures. *Phys. Rev. B* **1989**, *40*, 12359–12381. [CrossRef]
51. Cao, D.H.; Stoumpos, C.C.; Farha, O.K.; Hupp, J.T.; Kanatzidis, M.G. 2D Homologous Perovskites as Light-Absorbing Materials for Solar Cell Applications. *J. Am. Chem. Soc.* **2015**, *137*, 7843–7850. [CrossRef]

52. Ahmad, S.; Fu, P.; Yu, S.; Yang, Q.; Liu, X.; Wang, X.; Wang, X.; Guo, X.; Li, C. Dion-Jacobson phase 2D layered perovskites for solar cells with ultrahigh stability. *Joule* **2019**, *3*, 794. [CrossRef]
53. Ren, H.; Yu, S.; Chao, L.; Xia, Y.; Sun, Y.; Zuo, S.; Li, F.; Niu, T.; Yang, Y.; Ju, H.; et al. Efficient and stable Ruddlesden–Popper perovskite solar cell with tailored interlayer molecular interaction. *Nat. Photonics* **2020**, *14*, 154–163. [CrossRef]
54. Gan, Z.; Cheng, Y.; Chen, W.; Loh, K.P.; Jia, B.; Wen, X. Photophysics of 2D organic–inorganic hybrid lead halide perovskites: Progress, debates, and challenges. *Adv. Sci.* **2021**, *8*, 2001843. [CrossRef] [PubMed]
55. Bychkov, Y.A.; Rashba, E.I. Properties of a 2D Electron-Gas with Lifted Spectral Degeneracy. *JETP Lett.* **1984**, *39*, 78–81.
56. Nitta, J.; Akazaki, T.; Takayanagi, H.; Enoki, T. Gate Control of Spin-Orbit Interaction in an Inverted In0.53Ga0.47As/In0.52Al0.48As Heterostructure. *Phys. Rev. Lett.* **1997**, *78*, 1335–1338. [CrossRef]
57. King, P.D.; Hatch, R.C.; Bianchi, M.; Ovsyannikov, R.; Lupulescu, C.; Landolt, G.; Slomski, B.; Dil, J.H.; Guan, D.; Mi, J.L.; et al. Large Tunable Rashba Spin Splitting of a Two-Dimensional Electron Gas in Bi_2Se_3. *Phys. Rev. Lett.* **2011**, *107*, 096802. [CrossRef] [PubMed]
58. Lesne, E.; Fu, Y.; Oyarzun, S.; Rojas-Sánchez, J.C.; Vaz, D.C.; Naganuma, H.; Sicoli, G.; Attané, J.-P.; Jamet, M.; Jacquet, E.; et al. Highly efficient and tunable spin-to-charge conversion through Rashba coupling at oxide interfaces. *Nat. Mater.* **2016**, *15*, 1261–1266. [CrossRef] [PubMed]
59. Zhai, Y.; Baniya, S.; Zhang, C.; Li, J.; Haney, P.; Sheng, C.-X.; Ehrenfreund, E.; Vardeny, Z.V. Giant Rashba splitting in 2D organic-inorganic halide perovskites measured by transient spectroscopies. *Sci. Adv.* **2017**, *3*, e1700704. [CrossRef] [PubMed]
60. Todd, S.B.; Riley, D.B.; Binai-Motlagh, A.; Clegg, C.; Ramachandran, A.; March, S.A.; Hoffman, J.M.; Hill, I.G.; Stoumpos, C.C.; Kanatzidis, M.G.; et al. Detection of Rashba spin splitting in 2D organic-inorganic perovskite via precessional carrier spin relaxation. *APL Mater.* **2019**, *7*, 081116. [CrossRef]
61. Park, I.-H.; Zhang, Q.; Kwon, K.C.; Zhu, Z.; Yu, W.; Leng, K.; Giovanni, D.; Choi, H.S.; Abdelwahab, I.; Xu, Q.-H.; et al. Ferroelectricity and Rashba Effect in a Two-Dimensional Dion-Jacobson Hybrid Organic–Inorganic Perovskite. *J. Am. Chem. Soc.* **2019**, *141*, 15972–15976. [CrossRef]
62. Jana, M.K.; Song, R.; Liu, H.; Khanal, D.R.; Janke, S.M.; Zhao, R.; Liu, C.; Vardeny, Z.V.; Blum, V.; Mitzi, D.B. Organic-to-inorganic structural chirality transfer in a 2D hybrid perovskite and impact on Rashba-Dresselhaus spin-orbit coupling. *Nat. Commun.* **2020**, *11*, 4699. [CrossRef] [PubMed]
63. Kagdada, H.L.; Gupta, S.K.; Sahoo, S.; Singh, D.K. Mobility driven thermoelectric and optical properties of two-dimensional halide-based hybrid perovskites: Impact of organic cation rotation. *Phys. Chem. Chem. Phys.* **2022**, *24*, 8867–8880. [CrossRef] [PubMed]
64. Zhou, B.; Liang, L.; Ma, J.; Li, J.; Li, W.; Liu, Z.; Li, H.; Chen, R.; Li, D. Thermally Assisted Rashba Splitting and Circular Photogalvanic Effect in Aqueously Synthesized 2D Dion–Jacobson Perovskite Crystals. *Nano Lett.* **2021**, *21*, 4584–4591. [CrossRef] [PubMed]
65. Mitzi, D.B. Synthesis, structure and properties of organic-inorganic perovskites and related materials. *Prog. Inorg. Chem.* **1999**, *48*, 121.
66. Shao, Y.; Gao, W.; Yan, H.; Li, R.; Abdelwahab, I.; Chi, X.; Rogée, L.; Zhuang, L.; Fu, W.; Lau, S.P.; et al. Unlocking surface octahedral tilt in two-dimensional Ruddlesden-Popper perovskites. *Nat. Commun.* **2022**, *13*, 138. [CrossRef] [PubMed]
67. Singh, S.; Gong, W.; Stevens, C.E.; Hou, J.; Singh, A.; Zhang, H.; Anantharaman, S.B.; Mohite, A.D.; Hendrickson, J.R.; Yan, Q.; et al. Valley-Polarized Interlayer Excitons in 2D Chalcogenide–Halide Perovskite–van der Waals Heterostructures. *ACS Nano* **2023**, *17*, 7487–7497. [CrossRef] [PubMed]
68. Yin, J.; Maity, P.; Xu, L.; El-Zohry, A.M.; Li, H.; Bakr, O.M.; Brédas, J.-L.; Mohammed, O.F. Layer-Dependent Rashba Band Splitting in 2D Hybrid Perovskites. *Chem. Mater.* **2018**, *30*, 8538–8545. [CrossRef]
69. Liu, B.; Gao, H.; Meng, C.; Ye, H. The Rashba effect in two-dimensional hybrid perovskites: The impacts of halogens and surface ligands. *Phys. Chem. Chem. Phys.* **2022**, *24*, 27827–27835. [CrossRef] [PubMed]
70. Efros, A.L.; Brus, L.E. Nanocrystal Quantum Dots: From Discovery to Modern Development. *ACS Nano* **2021**, *15*, 6192–6210. [CrossRef]
71. Geiregat, P.; Rodá, C.; Tanghe, I.; Singh, S.; Di Giacomo, A.; Lebrun, D.; Grimaldi, G.; Maes, J.; Van Thourhout, D.; Moreels, I.; et al. Localization-limited exciton oscillator strength in colloidal CdSe nanoplatelets revealed by the optically induced stark effect. *Light Sci. Appl.* **2021**, *10*, 112. [CrossRef]
72. Swift, M.W.; Lyons, J.L.; Efros, A.L.; Sercel, P.C. Rashba exciton in a 2D perovskite quantum dot. *Nanoscale* **2021**, *13*, 16769. [CrossRef] [PubMed]
73. Li, J.; Guo, Z.; Qin, Y.; Liu, R.; He, Y.; Zhu, X.; Xu, F.; He, T. Rashba Effect and Spin-Dependent Excitonic Properties in Chiral Two-Dimensional/Three-Dimensional Composite Perovskite Films. *J. Phys. Chem. Lett.* **2023**, *14*, 11697–11703. [CrossRef] [PubMed]
74. Thouin, F.; Golub, L.; Lomakina, F.; Bel'kov, V.; Olbrich, P.; Stachel, S.; Caspers, I.; Griesbeck, M.; Kugler, M.; Hirmer, M.J.; et al. Phonon coherences reveal the polaronic character of excitons in two-dimensional lead halide perovskites. *Nat. Mater.* **2019**, *18*, 349–356. [CrossRef] [PubMed]
75. Lechner, V.; Golub, L.E.; Lomakina, F.; Bel'kov, V.V.; Olbrich, P.; Stachel, S.; Caspers, I.; Griesbeck, M.; Kugler, M.; Hirmer, M.J.; et al. Spin and orbital mechanisms of the magnetogyrotropic photogalvanic effects in $GaAs/Al_xGa_{1-x}As$ quantum well structures. *Phys. Rev. B* **2011**, *83*, 155313. [CrossRef]
76. Lee, J.S.; Schober, G.A.H.; Bahramy, M.S.; Murakawa, H.; Onose, Y.; Arita, R.; Nagaosa, N.; Tokura, Y. Optical Response of Relativistic Electrons in the Polar BiTeI Semiconductor. *Phys. Rev. Lett.* **2011**, *107*, 117401. [CrossRef] [PubMed]
77. Yuan, H.; Wang, X.; Lian, B.; Zhang, H.; Fang, X.; Shen, B.; Xu, G.; Xu, Y.; Zhang, S.-C.; Hwang, H.Y.; et al. Generation and electric control of spin–valley-coupled circular photogalvanic current in WSe_2. *Nat. Nanotechnol.* **2014**, *9*, 851–857. [CrossRef] [PubMed]

78. Dong, Y.; Zhang, Y.; Li, X.; Feng, Y.; Zhang, H.; Xu, J. Chiral perovskites: Promising materials toward next-generation optoelectronics. *Small* **2019**, *15*, 1902237. [CrossRef] [PubMed]
79. Dang, Y.; Liu, X.; Cao, B.; Tao, X. Chiral halide perovskite crystals for optoelectronic applications. *Matter* **2021**, *4*, 794–820. [CrossRef]
80. Ma, J.; Wang, H.; Li, D. Recent Progress of Chiral Perovskites: Materials, Synthesis, and Properties. *Adv. Mater.* **2021**, *33*, 2008785. [CrossRef]
81. Zhang, X.; Liu, X.; Li, L.; Ji, C.; Yao, Y.; Luo, J. Great Amplification of Circular Polarization Sensitivity via Heterostructure Engineering of a Chiral Two-Dimensional Hybrid Perovskite Crystal with a Three-Dimensional MAPbI$_3$ Crystal. *ACS Cent. Sci.* **2021**, *7*, 1261–1268. [CrossRef]
82. Huang, P.-J.; Taniguchi, K.; Shigefuji, M.; Kobayashi, T.; Matsubara, M.; Sasagawa, T.; Sato, H.; Miyasaka, H. Chirality-Dependent Circular Photogalvanic Effect in Enantiomorphic 2D Organic–Inorganic Hybrid Perovskites. *Adv. Mater.* **2021**, *33*, 2008611. [CrossRef] [PubMed]
83. Liu, T.; Shi, W.; Tang, W.; Liu, Z.; Schroeder, B.C.; Fenwick, O.; Fuchter, M.J. High Responsivity Circular Polarized Light Detectors based on Quasi Two-Dimensional Chiral Perovskite Films. *ACS Nano* **2022**, *16*, 2682–2689. [CrossRef] [PubMed]
84. Ward, M.D.; Shi, W.; Gasparini, N.; Nelson, J.; Wade, J.; Fuchter, M.J. Materials for optical, magnetic and electronic devices. *J. Mater. Chem. C* **2022**, *10*, 10452. [CrossRef] [PubMed]
85. Wang, J.; Lu, H.; Pan, X.; Xu, J.; Liu, H.; Liu, X.; Khanal, D.R.; Toney, M.F.; Beard, M.C.; Vardeny, Z.V. Spin-Dependent Photovoltaic and Photogalvanic Responses of Optoelectronic Devices Based on Chiral Two-Dimensional Hybrid Organic–Inorganic Perovskites. *ACS Nano* **2021**, *15*, 588–595. [CrossRef]
86. Fan, C.-C.; Han, X.-B.; Liang, B.-D.; Shi, C.; Miao, L.-P.; Chai, C.-Y.; Liu, C.-D.; Ye, Q.; Zhang, W. Chiral Rashba Ferroelectrics for Circularly Polarized Light Detection. *Adv. Mater.* **2022**, *34*, e2204119. [CrossRef]

Disclaimer/Publisher's Note: The statements, opinions and data contained in all publications are solely those of the individual author(s) and contributor(s) and not of MDPI and/or the editor(s). MDPI and/or the editor(s) disclaim responsibility for any injury to people or property resulting from any ideas, methods, instructions or products referred to in the content.

MDPI AG
Grosspeteranlage 5
4052 Basel
Switzerland
Tel.: +41 61 683 77 34

Nanomaterials Editorial Office
E-mail: nanomaterials@mdpi.com
www.mdpi.com/journal/nanomaterials

Disclaimer/Publisher's Note: The title and front matter of this reprint are at the discretion of the Guest Editor. The publisher is not responsible for their content or any associated concerns. The statements, opinions and data contained in all individual articles are solely those of the individual Editor and contributors and not of MDPI. MDPI disclaims responsibility for any injury to people or property resulting from any ideas, methods, instructions or products referred to in the content.

www.ingramcontent.com/pod-product-compliance
Lightning Source LLC
LaVergne TN
LVHW072357090526
838202LV00019B/2568